The Invisible Universe: A Quest to Discover Dark Matter

Johnny

TABLE OF CONTENTS

Part I

Introduction and Background

Introduction

The standard model (SM) of particle physics stands as one of the most successful theoretical frameworks in the history of science. From the unification of electromagnetic and weak forces to the prediction and subsequent discovery of the Higgs boson, the SM has provided a remarkably accurate and precise description of the fundamental particles and forces that govern the universe.

However, despite its successes, the SM is far from complete. Cosmological and astronomical observations have revealed that only 15% of the matter content in the universe are composed of visible matter that are well described by the SM. The majority of the universe consists of dark matter (DM), whose existence is inferred from the overwhelming number of astronomical and cosmological observations, such as measurements of galactic rotation curves, weak gravitational lensing observations of galaxy clusters, and the cosmic microwave background. However, all of these observations are through gravitational interactions, telling us little about the fundamental properties of DM. Deciphering its fundamental properties, including its mass, composition, and interactions with SM particles, remains one of the foremost open questions in basic science today.

For the past decades, many theoretically appealing hypotheses regarding the nature of DM have been proposed and actively pursued experimentally, ranging from massive compact halo objects to weakly interacting massive particles (WIMP). However, the lack of a direct DM detection, which excludes significant regions of the WIMP parameter space, and the absence of new physics at colliders have undermined the theoretical motivation behind these theoretical models. On the other hand, many other DM candidates have been proposed, among which hidden-sector DM and wave-like DM stand out as strongly motivated possibilities that have remained largely unexplored experimentally due to detector limitations. In light of this challenge, my doctoral research aims to unravel the possible particle- or wave-like nature of DM with two innovative and highly complementary and differentiated approaches: 1) to produce and detect hidden-sector particles with a novel reconstruction technique at the powerful Large Hadron Collider (LHC) and 2) to detect DM candidates from the galactic halo with advanced quantum sensing technology.

In the first approach, I exploit the high intensity and large center-of-mass energy collisions at the LHC to produce hidden-sector particles that are then detected efficiently by the Compact Muon Solenoid (CMS) with a unique technique that I developed. Many extensions of the SM predict the existence of neutral, weakly-coupled particles that have a long lifetime. These long-lived particles (LLPs) often provide striking displaced signatures in detectors, thus escaping the conventional searches for prompt particles and remaining largely unexplored at the LHC. I performed first searches [1, 2] at the LHC that use a muon detector as a sampling calorimeter to identify displaced showers produced by decays of LLPs. The searches are sensitive to LLPs decaying to final states including hadrons, taus, electrons, or photon, LLP masses as low as a few GeV, and is largely model-independent. The searches are enabled by the unique design of CMS muon detectors, composed of detector planes interleaved with the steel layers of the magnet flux-return yoke. Decays of LLPs in the muon detectors induce hadronic and electromagnetic showers, giving rise to a high hit multiplicity in localized detector regions that can be efficiently identified with a novel reconstruction technique. The searches yield competitive sensitivity for proper lifetime from 0.1 m to 1000 m with the full Run 2 dataset recorded at the LHC in 2016–2018. To extend the physics reach of this novel muon detector shower (MDS) signature, I demonstrated the model-independence of MDS and reinterpreted the search in a large number of LLP models [3, 4], illustrating its complementarity with many proposed and existing dedicated LLP experiments. Finally, I also contributed to the development of a new dedicated MDS trigger that has been successfully deployed in 2022 at the start of Run 3 of the LHC operations.

In parallel, to close the gap in DM discovery reach due to current limitations in detector sensitivity, I used for the first time, a novel quantum sensor, specifically the time-resolved, low-noise, single-photon sensitive superconducting nanowire single photon detectors (SNSPDs) to directly detect DM from our galaxy. I contributed to the development and characterization of SNSPDs for two entirely new experiments to directly detect axions via absorption and hidden-sector DM via electron scattering. The search for axions employs a novel broadband reflector technique with the Broadband Reflector Experiment for Axion Detection (BREAD). The BREAD experiment searches for axions or dark photons by using a unique parabolic mirror to focus axion or dark photon-converted photons to the SNSPDs. The SNSPDs allow us to be uniquely sensitive to 0.04–1 eV axions and dark photons that were not accessible before. I developed and built for the first time an SNSPD integration and characterization system to thoroughly benchmark the performance of state-of-the-art large area (mm^2) sensors towards the first stage dark photon pilot experiment.

On the other hand, by coupling the SNSPDs with gallium arsenide (GaAs), a bright cryogenic scintillator well matched to SNSPD detection, a prototype sensing system can be built as a basis of new direct DM detection experiments capable of extending the discovery to DM masses as low as 1 MeV. In particular, I built an x-ray calibration system based on the Compton scattering effect that measures the energy response of the sensing system. The work in this book lays the groundwork for a new era of DM direct detection experiment that leverages on the emerging field of quantum information science (QIS). With the breakthroughs in QIS revolutionizing detection techniques, new avenues emerge for probing the properties of DM across a wide range of parameter spaces that could offer new insights into the nature of our universe.

This book is divided into three parts. The remainder of Part I gives a brief introduction to the theoretical and phenomenological basis of the work in the book, including the foundations of the standard model in Chapter 2 and essential backgrounds for DM in Chapter 3. Part II describes the search for long-lived particles with the CMS muon detectors, consisting of four chapters. Chapter 4 provides an introduction to the LHC and the CMS experiment with an emphasis on the muon de-tectors. Chapter 5 describes the search performed using the LHC full Run 2 dataset. The reintepretation of the search using the endcap muon detectors is detailed in Chapter 6. The new trigger that has been commissioned in Run 3 and its prospects are described in Chapter 7. Finally, Part III details the progress of DM direct detec-tion experiments with SNSPDs by first introducing SNSPDs in Chapter 8 and then describing the search for axion with the BREAD experiment in Chapter 9 and the search for light hidden-sector DM with SNSPDs coupled to GaAs in Chapter 10.

The Standard Model of Particle Physics

2.1 Introduction

The Standard Model (SM) of particle physics describes all known elementary particles and their interactions through three of the four known fundamental forces (electromagnetic, weak, and strong interactions–excluding gravity). The theory was developed in stages throughout the latter half of the 20th century, including the work of many scientists worldwide. The current formulation was finalized in the mid-1970s, upon experimental confirmation of the existence of quarks. Since then, the discovery of the W and Z boson in 1983 [5–8], top quark in 1995 [9, 10], tau neutrino in 2000 [11], and finally the Higgs boson in 2012 [12, 13] have added further credence to the SM. The SM has been a highly successful theory that has withstand decades of high-precision testing.

The SM is a renormalizable quantum field theory, described by the $U(1)_Y \times SU(2)_L \times SU(3)_c$ gauge symmetry group, where Y is hypercharge, L is left-handedness, and c is color charge. Three fundamental forces in the SM arise due to the exchange for force carriers (spin-1 bosons) among the spin-$\frac{1}{2}$ fermions that make up matter. Each factor in the gauge symmetry group describes a fundamental force, represented by a gauge field, whose excitations are the gauge bosons that act as force carriers. The strong interaction has $SU(3)$ symmetry and is mediated by eight different types of gluons. The electroweak interaction has $U(1) \times SU(2)$ symmetry and is mediated by four bosons that mix to form the massive $W^{\pm}$ and Z bosons, and the massless photon (γ).

The matter fields are made up of fermions that can be further divided into two categories: quarks and leptons. There are six types of quarks: up (u), down (d), charm (c), strange (s), top (t), and bottom (b), and six types of leptons: electron (e), muon (μ), tau (τ), electron neutrino (ν_e), muon neutrino (ν_μ), and tau neutrino (ν_τ). The quarks interact through both strong and electroweak forces; the charged leptons interact only through the electroweak force; and the neutral leptons (neutrinos) interact solely through the weak force. Pairs of fermions can also be categorized into three generations or favors, as can be seen in Figure 2.1, where all known elementary particles are shown.

Finally, the Higgs Boson (H) is the only massive spin-0 scalar boson observed in nature. It is responsible for breaking the electroweak symmetry, giving rise to particle masses, which will be described in more detail in Section 2.2.

The interactions between the particles are summarized by the Lagrangian density:

$$\mathcal{L} = -\frac{1}{4}G^a_{\mu\nu}G^{a\mu\nu} - \frac{1}{4}W^a_{\mu\nu}W^{a\mu\nu} - \frac{1}{4}B_{\mu\nu}B^{\mu\nu}$$
$$+\bar{\psi}^i(i\gamma^\mu)(\mathcal{D}_\mu)_{ij}\psi^j$$
$$+\mathcal{L}_{\text{Higgs}} + \mathcal{L}_{\text{Yukawa}} \tag{2.1}$$

where:

- ψ is the fermion field;

- γ are the Dirac matrices;

- $\mathcal{D}_\mu$ is the gauge-covariant derivative, defined as $\mathcal{D}_\mu = \partial_\mu - ig_1 B_\mu Y - ig_2 W^a_\mu T^a - ig_3 G^a_\mu t^a$;

- $G^a_{\mu\nu}, W^a_{\mu\nu}, B_{\mu\nu}$ are the field strength tensors for SU(3)$_c$, SU(2)$_L$, U(1)$_Y$, respectively;

- a is the index of the generators, which runs from 1 to 8 for the gluon, and 1 to 3 for the weak $SU(2)_L$ group;

- μ, ν are the Lorentz vector indices;

- $\mathcal{L}_{\text{Higgs}}$ and $\mathcal{L}_{\text{Yukawa}}$ are the terms related to Higgs field and Yukawa interactions, respectively, which will be described further in Section 2.2.

2.2 Electroweak Symmetry Breaking

One feature of the non-abelian gauge theories, or Yang-Mills theories is that the gauge bosons and fermions are massless, since the gauge symmetry forbids explicit mass terms in the Lagrangian. Therefore, the Yang-Mills theory predicts massless spin-1 gauge bosons. Similarly, spontaneous symmetry breaking gives rise to Nambu-Goldstone bosons, and these additional bosons are massless spin-0 particles. However, massless weakly-interacting gauge bosons would lead to long-range forces, but they are only observed for electromagnetic and the corresponding massless photons. The gauge theories of the weak force needed a way to describe massive gauge bosons in order to be consistent.

In 1962, Philip Anderson first demonstrated that breaking gauge symmetries can lead to massive gauge bosons in non-relativistic field theory [15]. Shortly after, the

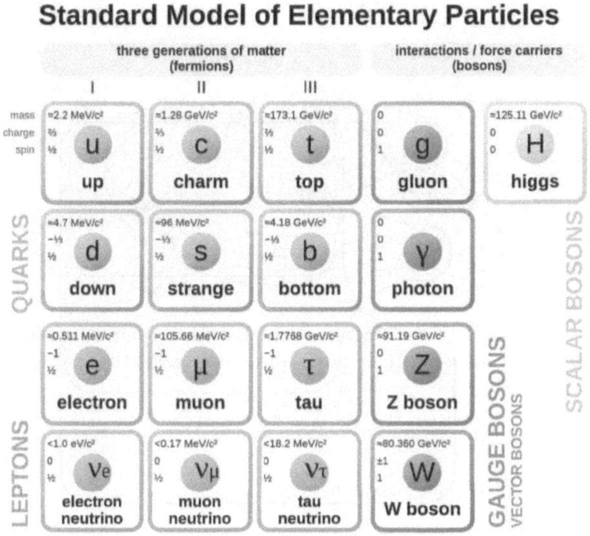

Standard Model of Elementary Particles

three generations of matter (fermions) — I, II, III

interactions / force carriers (bosons)

QUARKS

	mass	charge	spin
u (up)	≈2.2 MeV/c²	⅔	½
c (charm)	≈1.28 GeV/c²	⅔	½
t (top)	≈173.1 GeV/c²	⅔	½
d (down)	≈4.7 MeV/c²	−⅓	½
s (strange)	≈96 MeV/c²	−⅓	½
b (bottom)	≈4.18 GeV/c²	−⅓	½

LEPTONS

	mass	charge	spin
e (electron)	≈0.511 MeV/c²	−1	½
μ (muon)	≈105.66 MeV/c²	−1	½
τ (tau)	≈1.7768 GeV/c²	−1	½
νe (electron neutrino)	<1.0 eV/c²	0	½
νμ (muon neutrino)	<0.17 MeV/c²	0	½
ντ (tau neutrino)	<18.2 MeV/c²	0	½

GAUGE BOSONS / VECTOR BOSONS

	mass	charge	spin
g (gluon)	0	0	1
γ (photon)	0	0	1
Z (Z boson)	≈91.19 GeV/c²	0	1
W (W boson)	≈80.360 GeV/c²	±1	1

SCALAR BOSONS

	mass	charge	spin
H (higgs)	≈125.11 GeV/c²	0	0

Figure 2.1: Elementary particles in the Standard Model. Image reprinted from [14].

approach was extended to relativistic gauge theories by introducing a new scalar field that spontaneously breaks the symmetry group [16–18]. Finally, in 1967, Steven Weinberg and Abdus Salam independently incorporated this approach into the SM, as a gauge theory description of the electroweak force [19, 20]. In the SM, a scalar $SU(2)$ doublet field (the Higgs field) is introduced to spontaneously break the electroweak symmetry in vacuum to keep the structure of the gauge symmetry, while generating masses for the $W^{\pm}$ and Z gauge bosons and the fermions through Yukawa interactions with the Higgs field. This process is known as the Brout-Englert-Higgs mechanism, or simply the Higgs mechanism.

The Lagrangian term that describes the Higgs field can be written as:

$$\mathcal{L}_{\text{Higgs}} = (\mathcal{D}_{\mu}\Phi)(\mathcal{D}_{\mu}\Phi) - V(\Phi)\,, \quad V(\Phi) = -\mu^2\Phi^{\dagger}\Phi + \lambda(\Phi^{\dagger}\Phi)^2 \qquad (2.2)$$

where λ needs to be positive so that the potential ($V(\Phi)$) has a minimum value and Φ is a complex scalar field that is a $SU(2)_L$ doublet with weak hypercharge 1/2:

$$\Phi = \frac{1}{\sqrt{2}}\begin{pmatrix} \phi_1 \\ \phi_2 \end{pmatrix}, \qquad (2.3)$$

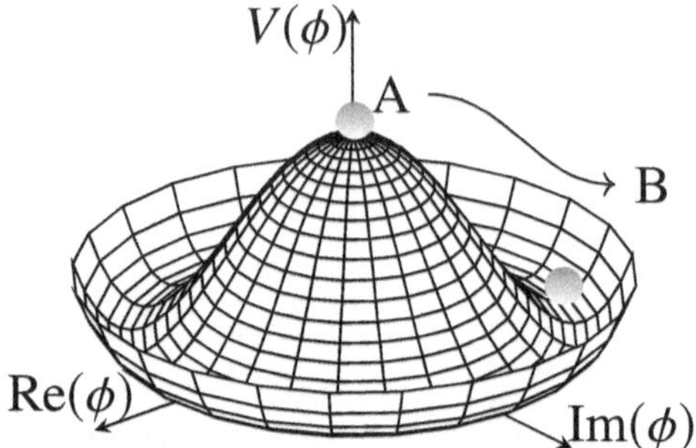

Figure 2.2: The shape of the "Mexican hat" potential, with the minimum of the potential occurring at a non-zero Φ value. Illustration created by TikZ code provided by Janosh Riebesell [21].

where ϕ_1 and ϕ_2 are complex scalar fields.

If $\mu^2 < 0$, then the potential will have a minimum at $|\Phi| = 0$, leading to a vacuum expectation value (vev) of 0. In this case, the gauge transformation is still invariant, resulting in no additional mass term required for the $W^\pm$ and Z bosons and fermions. Therefore, μ^2 has to be greater than zero. When $\mu^2 > 0$, then the potential will have a "Mexican hat" shape, as shown in Figure 2.2. The minimum value of the potential will occur at:

$$\Phi^\dagger \Phi = \frac{\mu^2}{2\lambda} = \frac{v^2}{2} \tag{2.4}$$

Therefore, Φ acquires a non-zero vev (v) that is not $SU(2)_L \times U(1)_Y$ invariant.

There are many solutions to Equation 2.4. One can choose a particular minimum, where $\phi_1 = 0$ and ϕ_2 is a real scalar field:

$$\phi_{\min} = \frac{1}{\sqrt{2}} \begin{pmatrix} 0 \\ v \end{pmatrix}, \tag{2.5}$$

If we expand the potential around its minimum:

$$\Phi = \frac{1}{\sqrt{2}} \begin{pmatrix} 0 \\ v + H(x) \end{pmatrix}, \tag{2.6}$$

and rewrite the Lagrangian terms associated to the potential:

$$-V(\phi) = +\mu^2 \Phi^\dagger \Phi - \lambda(\Phi^\dagger \Phi)^2 = -\lambda v^2 H^2 - \lambda v H^3 - \frac{\lambda}{4} H^4 + \text{const} \tag{2.7}$$

As shown in the above equation, the symmetry breaking mechanism give rise to a physical spin-0 boson, the Higgs boson, represented by the $H(x)$ real scalar field. From the equation, the mass of the Higgs boson at tree-level $m_H = \sqrt{2\lambda}v$ can be obtained from the first term and the second and third represent the triple Higgs and quartic Higgs interaction vertices, respectively.

In addition to a giving rise to a new physical spin-0 boson, electroweak symmetry breaking is also responsible for generating the masses of the gauge bosons in the SM. By evaluating the covariant derivative in Equation 2.2 on the Higgs field. The kinetic terms can be written as:

$$
\begin{aligned}
|(\mathcal{D}_\mu \Phi)|^2 &= \frac{1}{2} \begin{pmatrix} \partial_\mu - \frac{i}{2}(g_2 W_\mu^3 + g_1 B_\mu) & -\frac{ig_2}{2}(W_\mu^1 - iW_\mu^2) \\ -\frac{ig_2}{2}(W_\mu^1 - iW_\mu^2) & \partial_\mu + \frac{i}{2}(g_2 W_\mu^3 - g_1 B_\mu) \end{pmatrix} \begin{pmatrix} 0 \\ v \end{pmatrix} \\
&= \frac{1}{8}(g_2^2 v^2 |W_\mu^1 + iW_\mu^2|^2 + v^2 |g_2 W_\mu^3 - g_1 B_\mu|^2)
\end{aligned}
\tag{2.8}
$$

where we can define four field combinations that correspond to the physical $W^\pm$, Z, and photon fields:

$$
\begin{aligned}
W_\mu^\pm &= \frac{W_\mu^1 \mp iW_\mu^2}{\sqrt{2}}, & m_W &= \frac{1}{2}v g_2 \\
Z_\mu &= \frac{g_2 W_\mu^3 - g_1 B_\mu}{\sqrt{g_1^2 + g_2^2}}, & m_Z &= \frac{1}{2}v\sqrt{g_1^2 + g_2^2} \\
A_\mu &= \frac{g_2 W_\mu^3 + g_1 B_\mu}{\sqrt{g_1^2 + g_2^2}}, & m_\gamma &= 0
\end{aligned}
\tag{2.9}
$$

This allow us to re-write Equation 2.8 to:

$$
|(\mathcal{D}_\mu \Phi)|^2 = m_W^2 W_\mu^+ W^{-\mu} + \frac{1}{2}m_Z^2 Z_\mu Z^\mu + \frac{1}{2}m_\gamma^2 A_\mu A^\mu
\tag{2.10}
$$

Thus, the $W^\pm$ and Z bosons have acquired mass and the photons remain massless.

Additionally, as proposed by Steven Weinberg [19], fermions acquire mass through interaction with the Φ field that has a non-zero vev. The Yukawa terms are added to the Lagrangian for each generation:

$$
\mathcal{L}_{\text{Yukawa}} = -y_e \bar{L}_L \Phi e_R - y_u \bar{Q}_L \tilde{\Phi} u_R - y_d \bar{Q}_L \Phi d_R + (\text{h.c.})
\tag{2.11}
$$

Then we can identify the fermion masses as:

$$
m_e = \frac{y_e v}{\sqrt{2}}, \qquad m_u = \frac{y_u v}{\sqrt{2}}, \qquad m_d = \frac{y_d v}{\sqrt{2}}.
\tag{2.12}
$$

2.3 Limitations of the SM

The observation of the Higgs boson in 2012 completed the last missing piece in the SM. Although the SM has been highly successful in its accurate predictions, it is not a complete theory of our universe. Notably, the SM does not describe the fourth fundamental force, gravity and does not predict the neutrino mass. More importantly, the SM is a theory that only describes the visible matter, which only makes up a mere 5% of the energy content of the universe in the present day, as shown in Figure 2.3. The remaining 95% of the universe is made of dark energy and dark matter (DM) that don't interact much with visible matter, and thus have been difficult to study.

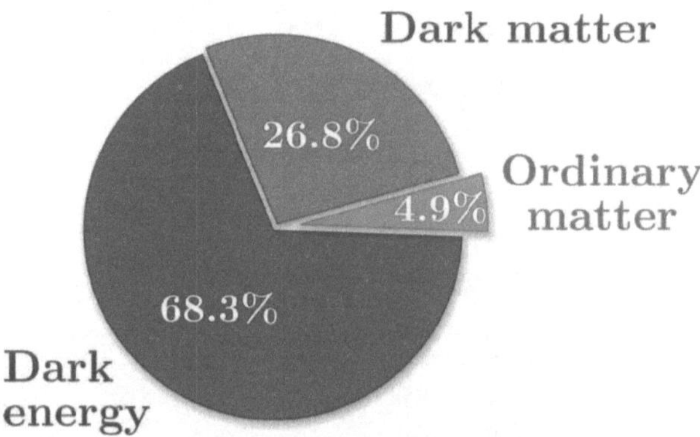

Figure 2.3: Energy composition of the universe from the high-precision measurement of the cosmic microwave background from the Planck satellite [22]. Image reprinted from [23].

The energy composition of our universe and the existence of dark energy have been inferred from cosmological observations of inflation and the cosmic microwave background. However, since we have not been able to directly observe dark energy in a controlled laboratory environment, we have little understanding of its nature.

About 85% of the matter in the universe is dark matter, which we know interacts gravitationally, but not electromagnetically, hence the name "dark matter". Similar to dark energy, we have not been able to directly detect them in the laboratory, so the only understanding comes astronomical and cosmological observations of large bodies in the universe. However, since DM can create relatively smaller scale effects within galaxies, precise astronomical observations have allowed us to have a better understanding of the requirements on its identity.

The following chapter will briefly summarize the evidence that we have for the existence of DM and introduce a few theoretically motivated DM candidates that are sought after by experiments and are the focus of this book.

3.1 Evidence for Dark Matter

3.1.1 Galaxy Rotation Curves

The first experimental evidence that the universe contains additional matter was the observations of the Coma Cluster by Franz Zwicky in 1933. Zwicky estimated the mass of the cluster with two methods and found them to be inconsistent. Zwicky first measured the mass by using the standard mass-to-light ratio. He then used a second method, where he measured the motion of the individual galaxies orbiting in the cluster by measuring the doppler shifts in their emission spectra. Using the virial theorem, he calculated the total mass of the cluster:

$$\frac{GM}{r} \sim v^2 \tag{3.1}$$

where G is the gravitational constant, M is the mass inside a galaxy's orbit, and v is the velocity of the galaxy relative to the center of mass of the cluster. He found out that the first method yielded a cluster mass that is only 2% of the mass measured from the motion of the galaxies. This discrepancy indicated that most of the mass (98%) in the cluster was not luminous and did not interact electromagnetically, and was thus dubbed the name "dark matter" (DM).

In the 1970s, Vera Rubin led a team to quantify the missing mass more precisely [24]. They observed 21 diverse galaxies and measured the velocities of the stars as a function of the radius to the galactic center. They showed that most galaxies contain about six times as much DM as visible matter and the mass was distributed much farther from the galactic center than the majority of the visible matter. This effect is shown in Figure 3.1 for spiral galaxy NGC 3198.

3.1.2 Gravitational Lensing

One direct consequence of general relativity is that the geometry of space-time is modified by massive objects. As a result, massive objects (in this case, DM in clusters of galaxies) that lie between a more distant source and an observe would act as a lens to bend the light from the source, and the amount of bending is proportional to the mass of the object. This bending effect on light is called gravitational lensing.

Strong lensing effect is observed when an obvious distortion of the background galaxy is seen. By measuring the distortion geometry, the mass of the intervening cluster can be calculate, thus the distribution of DM [26] can be mapped. Smaller

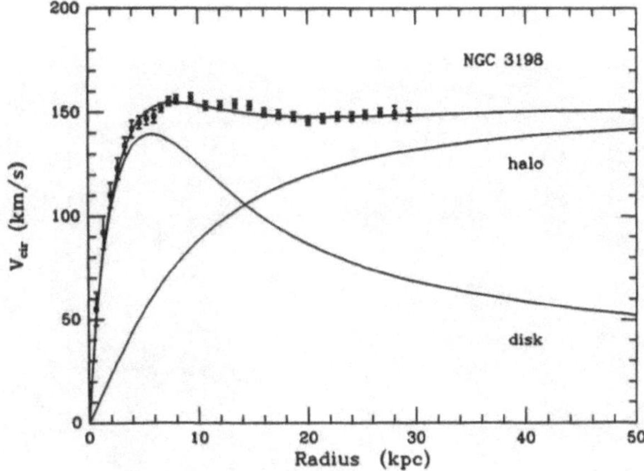

Figure 3.1: The measured rotation curve and fits of disk and halo of spiral galaxy NGC 3198 [25] are shown. The data points are measurements of the velocity of galactic hydrogen gas. The disk curve shows a model for the luminous matter in the galactic disk that is mostly concentrated in the galactic center. The halo curve fits to the additional DM, which is distributed farther from the galactic center, causing the outer stars to orbit faster.

lensing effects are usually more common and often more useful in mapping the DM distribution. Weak gravitation lensing surveys a vast number of galaxies and perform statistical analysis on the deformation of adjacent background galaxies.

One particular strong piece of evidence for DM comes from the observation of gravitational lensing in the Bullet Cluster. The Bullet Cluster is formed by a small cluster (the bullet) colliding with a larger cluster, as shown in Figure 3.2. During the collision, compact objects like stars and galaxies, being relatively sparse and far apart, largely pass by each other without any interaction. However, the intergalactic gas and plasma, that dominate the baryonic matter component of the cluster, are distributed more widely and interact during the collision. As a result of the interactions, the gas and plasma emit x-rays that are observed by the Chandra X-ray Observatory, that produced a map of the majority of the baryonic or visible mass in the cluster. On the other hand, weak gravitational lensing of background galaxies through the Bullet Cluster was performed that resulted in a very different mass distribution. In fact, the result of weak gravitation lensing concluded that the majority of the clusters's mass is non-baryonic in nature [27].

Figure 3.2: A composite image of the Bullet Cluster is shown [28]. The Bullet Cluster is formed by a pair of galaxy clusters colliding head on. The smaller cluster passes through the larger one from left to right in the image, like a bullet. The optical image from the Magellan and the Hubble Space Telescope shows galaxies in orange and white in the background. Hot gas (pink), which contains the bulk of the baryonic or visible matter in the cluster, is shown by the Chandra X-ray image. Gravitational lensing, the distortion of background images by mass in the cluster, reveals the mass of the cluster is dominated by DM (blue).

3.1.3   The Cosmic Microwave Background

A compelling evidence that the DM existed since the early universe comes from its effect on the cosmic microwave background (CMB), which is one of the earliest probe we have of our nascent universe. In the early universe, baryonic matter was ionized and interacted strongly with radiation via Thomson scattering. DM does not interact with radiation, but affects CMB through gravitational interaction and its effect on the density and velocity of baryonic matter. Therefore, baryonic and dark matter perturbations evolve differently over time and leave different signatures on the CMB.

The CMB is mostly isotropic and uniform, with a blackbody temperature of 2.73K, but it does have a temperature fluctuation of about 1 part in 10^5. The anisotropies can be decomposed into an angular power spectrum, where a series of acoustic peaks at almost-equal spacings but different heights are observed. The most precise power spectrum measured today is performed by the Planck satellite, as shown in Figure 3.3. The series of peaks are constrained by the cosmological parameters. The first peak mostly shows the baryonic matter density and the third peak is related to the DM density. The most recent result from Planck can be well fitted by the standard cosmological model [22], with measured relative densities to be 68.3%, 26.8%, 4.9% for dark energy, dark matter, and baryonic matter, respectively.

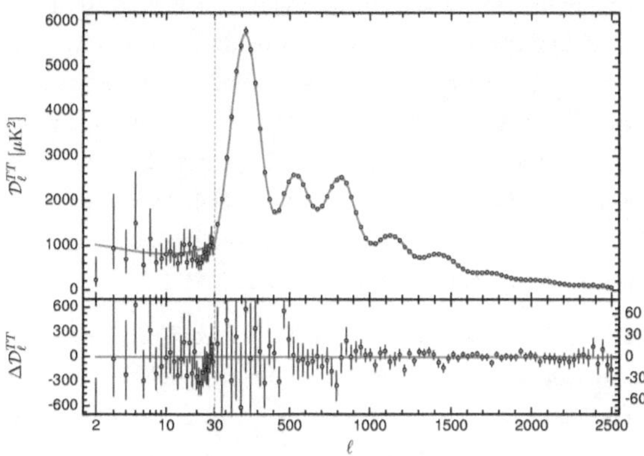

Figure 3.3: The CMB anisotropy spectrum measured by the Planck satellite [22]. The relative heights of the peaks in the power spectrum determine the cosmological parameters.

3.2 Dark Matter Properties

The previous section gave a brief overview of the substantial evidence for an additional matter component that has minimal interaction with SM particles. The evidence for DM is one of strongest indications for physics beyond the SM. Consequently, there have been numerous efforts to directly detect DM to understand its properties. However, other than indirect astronomical observations via gravitational interactions, DM has still yet to be directly detected in the laboratory.

Based on our astronomical observations of DM for decades, we can deduce some of the requirements of the properties of DM candidates.

Electric charge: Constraints from the CMB and large-scale structures require the

DM to be electrically neutral. If the DM were charged or milli-charged, it would impact the baryon-photon plasma during recombination, with its effect seen in the measured baryon acoustic peak structure.

Self-interactions: Observations of merging clusters, such as the Bullet Cluster [29], constrain the level of DM-DM self interactions. During these colliding events, the DM from the two clusters must interact minimally to pass through during the collisions, based on observations of gravitational lensing.

Mass: The mass of the DM extends many orders of magnitude from $\sim 10^{-22}$ eV $-$ $-5M_\odot$ [30]. The lower mass limit is limited by the Compton wavelength of the DM, $\lambda_{DM} = \hbar c/m_{DM}$. If the Compton wavelength is too large, it might erase small-scale structures that are observed in CMB [31] and measurements of galaxy luminosity [32–34]. The upper limit in mass is limited by the mass of the structures that the DM is immersed in, such as galactic disks, globular clusters, and individual small galaxies. The most stringent limits are derived from wide halo binaries [35] and the stability of the star cluster within Eridanus II [36].

Temperature: DM candidates are expected to be cold or non-relativistic in the early universe during structure formation. Being non-relativistic allows the DM candidates to be pulled by the gravitational wells of galaxies and clusters, as we observe today. The cold dark matter theory has been favored by cosmological observations, including the cosmic microwave background. The fact that DM has to be cold also invalidates neutrino, given its minimal SM interactions, as a DM candidate, since neutrino is relativistic.

Fermion or boson: The local DM energy density in the galactic halo can be extracted from the measured rotation curve. The most recent measurement indicates that the local DM energy density is $\rho_{DM} \approx 0.2 - 0.6 \, \text{GeV}/\text{cm}^3$ [30]. There is some controversy over the correct value to use for DM searches. Most DM experiments use $\rho_{DM} = 0.45 \, \text{GeV}/\text{cm}^3$, which is the value we use in the DM experiments in Chapter 9 and 10. The local DM energy density, gives us a relationship between the DM mass and the number density: lighter DM must have a higher number density, while heavier DM have lower density. For masses below $\sim 100 \, \text{eV}$, fermionic DM is limited by the Pauli Exclusion Principle and can no longer pack together tightly enough to maintain the required number density. Therefore, DM candidates lighter than that must be bosonic [37–39].

Lifetime: The DM lifetime must be long compared to cosmological timescales [40].

3.3 Dark Matter Candidates

There are a large number of theories that hypothesize the nature of DM. Many DM models pursue DM candidates that also solve other open issues in the SM, for example weakly interacting massive particles (WIMPs) that also address the hierarchy problem, axions that solve the strong CP problem, sterile neutrinos that are connected to problems of neutrino mass and mixing. Figure 3.4 catalogs a number of possible DM candidates, organized by their masses.

For the past decades, experimental efforts have focused primarily on searching WIMPs that are theoretically appealing. However, the lack of a direct DM detection, which excludes significant regions of the WIMP parameter space, and the absence of new physics at colliders have undermined the theoretical motivation behind it. On the other hand, many other DM candidates have been proposed, among which hidden-sector DM and wave-like DM stand out as strongly motivated possibilities and are the focus of this book.

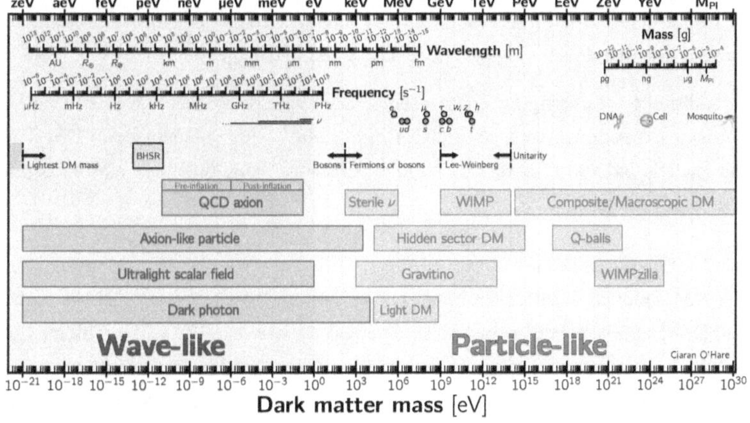

Figure 3.4: A number of DM particle hypotheses with their range of masses are shown. The x-axis is broken into two major classifications of DM particle hypotheses, particle-like DM and wave-like DM. Image is created by Ciaran O'Hare [41].

3.3.1 Hidden Sector Dark Matter

The absence of any conclusive signals from DM as a particle has motivated that the DM is charged under a new "hidden" sector [42], including mirror DM and more recently models in the context of neutral naturalness. The top-down motivation for hidden-sector DM comes from string theory [43]. The mass scale for hidden-sector DM is relatively broad, ranging from keV to TeV scale [44]. While electroweak-scale hidden sectors are theoretically motivated, light (sub-GeV) hidden sectors also have strong theoretical underpinnings and offer novel detection avenues and opportunities.

The phenomenology of hidden-sector DM depends primarily on the nature of the force and its force carrier. The dark sector couples to the visible sector through renormalizable "portals", including the Higgs portal, dark photon portal, and neutrino portal. In Chapter 5, a sensitive probe of the hidden sector through the Higgs portal, by searching for long-lived particle (LLPs) signatures with the CMS detector is presented. The unique signature from LLPs allowed us to probe weak couplings and light masses below GeV, that was thought to be inaccessible at the LHC. Additionally, in Chapter 6, the result of the search is re-interpreted to other hidden sector portals, including the dark photon and neutrino portals, and other models, including the axion-like particles and inelastic DM models. Improvement of the search in current ongoing data taking periods is presented in Chapter 7.

In addition to searching for hidden-sector DM that are produced in proton-proton collisions, a complementary search for light (MeV– GeV) hidden-sector DM from the galactic halo with direct detection experiments using quantum sensors is presented in Chapter 10.

3.3.2 Wave-like Dark Matter

For DM candidates with even lower masses, the number density of DM is higher and DM candidates behave more like classical wave. Wave-like DM candidates are usually bosons, and so they can be (pseudo-)scalar or (axial-)vector, with each possibility carrying a different set of interactions and detection technique and signatures. The most compelling wave-like DM candidates are pseudo-scalar QCD axion that was originally proposed to solve the strong CP problem and axion-like particles and vector dark (or hidden) photons that also arise in hidden sector models. A search for axions (QCD axion and axion-like particles) and dark photons with a novel broadband technique, called the BREAD experiment, is presented in Chapter 9 of this book.

Part II

Search for Long-lived Particles with the CMS Muon Detectors

The CMS Experiment at the LHC

This chapter presents an overview of the experimental facilities that are used to search for long-lived hidden-sector particles. Section 4.1 introduces the Large Hadron Collider (LHC), a hadron accelerator and collider that produces the high energy hadron collision events. Section 4.2 describes the Compact Muon Solenoid (CMS) detector that detects and collects LHC collisions, trigger system used to select interesting events, the data acquisition system, and event reconstruction techniques. An extensive and complete review of the LHC and CMS can be found in [45] and [46], respectively.

4.1 The Large Hadron Collider

The LHC is a 26.7 km two-ring-superconducting-hadron accelerator and collider located underground at the French-Swiss border near Geneva. With its designed maximum centre-of-mass energy of 14 TeV and an unprecedented luminosity of $10^{34}\,\mathrm{cm}^{-2}\,\mathrm{s}^{-1}$ for proton-proton collisions, the LHC is the world's most powerful particle collider ever built. Additionally, the LHC can also collide heavy ions (Pb) with a centre-of-mass energy of 5.5 TeV and luminosity of $10^{27}\,\mathrm{cm}^{-2}\,\mathrm{s}^{-1}$. The LHC was built to discover and study the Higgs boson properties or alternative mechanism responsible for electroweak symmetry breaking and search for beyond the SM (BSM) phenomena above the electroweak scale up to the TeV scale.

To accelerate the protons to a beam energy of 7 TeV needed in the LHC, a chain of small accelerators are need. The layout of the CERN accelerator complex is shown in Figure 4.1. The proton sources are produced from a bottle of hydrogen gas that is injected into a vacuum chamber, where strong electric field strips away the electrons. The resulting protons are then injected into a chain of four accelerators that group the protons to bunches of about 10^{11} protons with bunch spacing of 25 ns and boost the proton energy to 450 GeV before they are injected into the LHC. The accelerator chain is composed of a linear accelerator (Linac2), the Proton Synchrotron Booster (PSB), the Proton Synchrotron (PS), and the Super Proton Synchrotron (SPS) that accelerate the protons sequentially to 50 MeV, 1.4 GeV, 25 GeV, and 450 GeV, respectively. The 450 GeV proton beans are then injected into the LHC as two counter-rotating beams.

Inside the LHC, the proton beams travel in opposite directions in two separate ultrahigh vacuum (10^{-11} mbar) beam pipes. The beams are guided in the accelerator

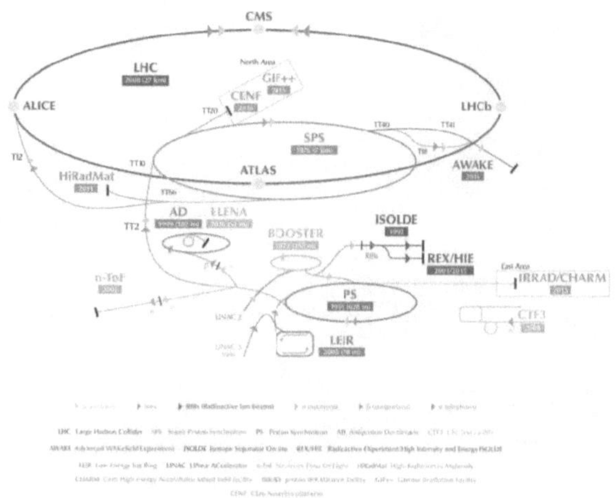

Figure 4.1: The full CERN accelerator complex. [47]

ring by a 8.33 T magnetic field provided by 1232 superconducting dipole magnets that bend the particle's trajectory and 392 quadruple magnetics that focus the beam. All of the LHC magnets are based on niobium-titanium Rutherford cables that are cooled to a temperature below 2 K with superfluid helium to operate at fields above 8 T. As a cost-effective solution, the LHC re-used the tunnel that was initially constructed between 1984 and 1989 for the Large Electron Positron (LEP) collider.. Therefore, the LHC design was influenced by the space-constraint of the existing tunnel, which is composed of eight crossing points and eight long straight sections for radio-frequency (RF) cavities. The tunnel in the arcs has a finished internal diameter of 3.7 m, making it extremely difficult to install two completely separate proton rings with separate magnet systems. Therefore, the twin-bore magnet design proposed by John Blewett in 1971 was adopted, where two sets of coils and beam pipes are accommodated in one mechanical structure and cryostat, with magnetic flux circulating in the opposite direction through the two channels. The cross section of the dipole is shown in Figure. 4.2. The proton beams are accelerated by 16 superconducting radio-frequency cavities in the LHC. Each proton gains 485 keV energy during one revolution around the LHC, which means 1.35×10^7 turns are needed to ramp up the energy from 450 GeV to 7 TeV, corresponding to a ramp up time of about 20 minutes.

LHC DIPOLE : STANDARD CROSS-SECTION

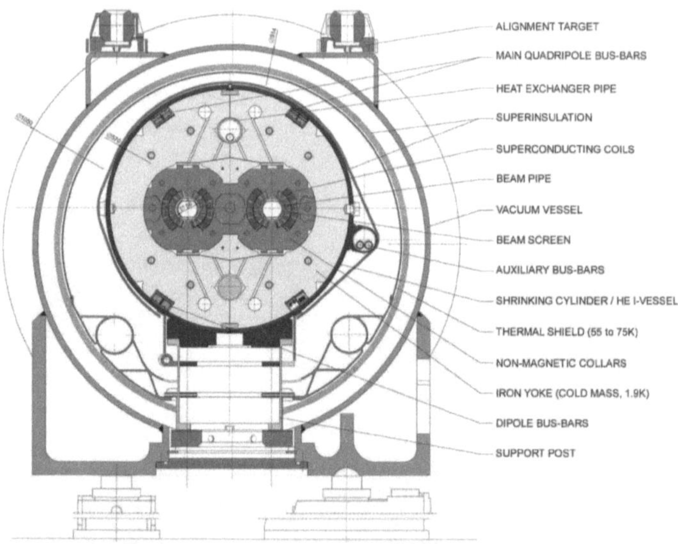

Figure 4.2: The cross section of the LHC dipole magnet [45].

To be able to explore rare processes in both the SM and BSM, it is crucial to produce a large number of events at the LHC, which is proportional to the luminosity that the LHC can deliver. The number of events (N) that are expected for a particular physics process is given by the product of the process cross section (σ) and the time integral of the instantaneous luminosity ($\mathcal{L}$):

$$N = \sigma \int \mathcal{L}(t)\, dt \tag{4.1}$$

The instantaneous luminosity depends on the LHC beam parameters and can be written for a Gaussian beam distribution [45]:

$$\mathcal{L} = \frac{N_b^2 n_b f_{\mathrm{rev}} \gamma_r}{4\pi \epsilon_n \beta*} F \tag{4.2}$$

where the definitions and the design values of the parameters that correspond to the nominal LHC peak luminosity of $10^{34}\,\mathrm{cm}^{-2}\,\mathrm{s}^{-1}$ are:

- N_b is the number of particles per bunch: 1.15×10^{11}.

- n_b is the number of bunches per beam: 2808 (Given 25 ns bunch spacings, there are 3564 bunch places in total. Empty bunches are kept to allow time for beam dump).

- f_{rev} is the revolution frequency: 11.245 kHz

- γ_r is the relativistic gamma factor: 7461 for 7 TeV protons.

- ϵ_n is the normalized transverse beam emittance: 3.75 μm

- $\beta*$ is the beta function at the collision point: 0.55 m

- F is the geometric luminosity reduction factor due to the crossing angle at the IP and can be written as $F = \left(1 + \left(\frac{\theta_c \sigma_z}{2\sigma*}\right)^2\right)^{-1/2}$. F is a function of the crossing angle (θ_c = 285 μrad), the bunch length in z (σ_z = 7.55 cm), and the transverse beam size ($\sigma*$ = 16.7 μm), assuming two equal round beams with $\sigma_z \ll \beta$. F is about 0.84.

The LHC luminosity decays over time from the peak value due to the degradation of the intensity and emittance of the beams. The main cause of the luminosity decay is due to beam loss from pp collisions that happen at the two high luminosity experiments reducing the intensity of the beam and the overall beam quality. Other contributions to beam loss come from the degradation of beam quality due to interactions of particles with residual gas in the beam pipe, beam-beam interactions among themselves, and noise in the radio-frequency system. As a result, the luminosity lifetime of the beam is about 15 hours, out of which about 10 hours are usually used for physics runs. The amount of data available is determined by the time integral of the luminosity. Assuming a turnaround time of 7 hours, optimum physics run time of 12 hours per fill, and 150 days of operation, one would obtain a maximum total luminosity per year of about 60 fb^{-1}.

Beams are collided in the LHC by crossing the beam paths with a small crossing angle (θ_c) at four collision points, where the four LHC experiments are located. The two high-luminosity, general-purpose detectors CMS and ATLAS are located at point 1 and point 5, at the opposite ends on the LHC circumference, respectively. There are also two specialized experiments, LHCb, located at point 8, is designed to study the charm and bottom quarks, and ALICE, located at point 2, is designed to study heavy-ion physics.

The rest of Part II in this book uses data collected by the CMS experiment, so proton-proton collision recorded by the CMS detector will be discussed in more detail. The LHC started the first physics data delivery in the spring of 2010 with a center-of-mass energy of 7 TeV, during a period called Run 1. During Run 1 (2010-2011), CMS collected 6 fb^{-1} of integrated luminosity. In 2012, the center-of-mass energy was increased to 8 TeV and 23.3 fb^{-1} of data were collected. From the beginning of 2013, the LHC was shut down for two years to prepare for Run 2, when both the center-of-mass energy and luminosity were significantly increased. Run 2 of the LHC restarted in 2015 at a center-of-mass energy of 13 TeV and lasted until end of 2018. The integrated luminosity recorded by CMS was 2.26, 36.3, 41.5, and 59.7 fb^{-1} for 2015, 2016, 2017, and 2018, respectively. The data collected in 2015 are usually not included in analyses, including the analysis presented in this book, since the 2015 data only adds 2% to the rest of the Run 2 dataset, but adding it would require significant time and resources to generate separate Monte Carlo simulations, calculate dedicated calibrations, and monitor detector performance and event reconstructions.

The peak luminosity achieved in 2016-2018 were about $1.5 - 2 \times 10^{34}\,\mathrm{cm}^{-2}\,\mathrm{s}^{-1}$, exceeding the design value of $1 \times 10^{34}\,\mathrm{cm}^{-2}\,\mathrm{s}^{-1}$. However, the center-of-mass energy was only at 13 TeV during Run 2, slightly below the designed 14 TeV. Despite that all of the LHC magnets were commissioned to a collision energy of 14 TeV, some of the dipole magnets have lower memory than expected, demanding larger number of quenches to reach the nominal field. Retraining the magnets to 13 TeV required relatively short amount of time, while retraining to 14 TeV would take significantly longer, taking time away from data taking. Therefore, to prioritize the potential to discovery new physics timely, Run 2 was operated at 13 TeV, which is still the highest energy that has been achieved at the time. After a second shutdown that started in 2019, LHC has restarted with an even higher center-of-mass energy of 13.6 TeV in summer 2022 with the goal of accumulating 300 fb^{-1} by the end of 2025. A summary plot of the integrated luminosity delivered to CMS is shown in Figure 4.3. The analysis presented in Chapter 5 uses the 13 TeV data collected from 2016-2018. A new dedicated trigger that would significantly improve the sensitivity of the Run 2 analysis has been commissioned for Run 3 and its prospects are discussed in Chapter 7.

When two proton bunches cross each other in the LHC, very often more than one pair of protons interact. Along with the hardest interaction that usually causes the

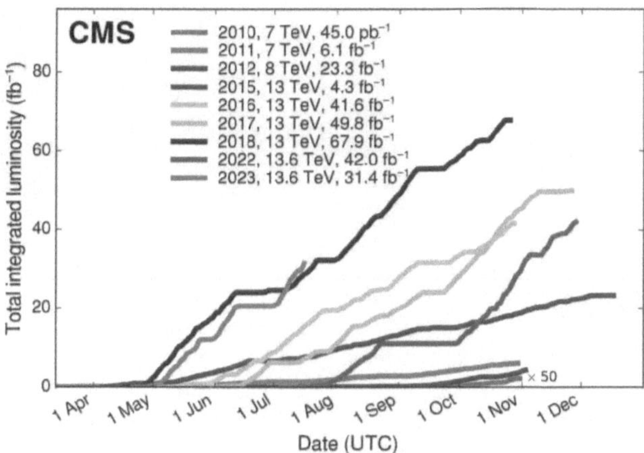

Figure 4.3: Cumulative luminosity versus day delivered to CMS during stable beams for pp collisions [48].

detector to trigger the data acquisition, the event is accompanied by several softer collisions between other protons in the bunches. The number of simultaneous pp interactions occurring in the same bunch crossing, called pileup, can be estimated from:

$$N_{pileup} = \frac{\sigma_{\text{inel}} L}{n_b f_{\text{rev}}} \tag{4.3}$$

where σ_{inel} is the inelastic pp cross section which is 80 mb at 13 TeV [49], L is the luminosity, n_b is the number of proton per bunch crossing, and f_{rev} is the revolution frequency. Given the cross section and the LHC beam parameters, the expected number of pileup per bunch crossing is about 50 for the LHC. Since the luminosity varies with time, the average number of pileup per bunch crossing also varies. The distributions of the mean number of interactions per crossing for each year during Run 2 is shown in Figure. 4.4. Pileup interactions produce a large number of soft particle that can be challenging to be distinguished from the hard scattering event of interest. Therefore, the ability to mitigate pileup to be able to reconstruct interesting event is crucial to operating the LHC at high luminosity.

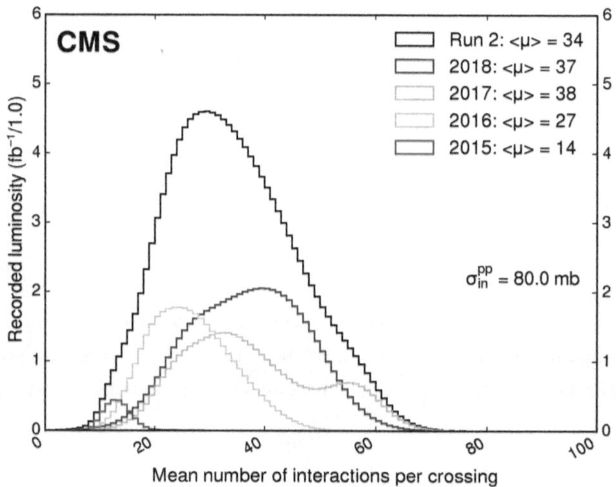

Figure 4.4: Distributions of the average number of interactions per crossing (pileup) for pp collisions in 2015 (purple), 2016 (orange), 2017 (light blue), and 2018 (navy blue) [48].

4.2 The Compact Muon Solenoid Detector

The Compact Muon Solenoid (CMS) detector is one of the two general purpose detectors at the LHC. It was designed to study proton-proton (and lead-lead) collisions at a center-of-mass energy of 14 TeV (5.5 TeV), with a design luminosity of $10^{34}\,\mathrm{cm}^{-2}\,\mathrm{s}^{-1}$ ($10^{27}\,\mathrm{cm}^{-2}\,\mathrm{s}^{-1}$). The detector is installed at Point 5 of the LHC, about 100 meters underground close to the French village of Cessy, between Lake Geneva and the Jura mountains. The prime motivation of the LHC is to study the nature of electroweak symmetry breaking and search for evidence of new phenomena at energy scales above about 1 TeV that could pave the way toward a unified theory beyond the Standard Model. Additionally, the LHC will also provide high-energy heavy-ion beams, allowing us to further extend the study of QCD matter under extreme conditions of temperature, density, and parton momentum fraction (low-x).

The seven-fold increase in energy and a hundred-fold increase in integrated luminosity at the LHC over the Tevatron [50], the previous circular hadron collider operating at Fermilab until 2011, lead to formidable experimental challenges to the particle detectors. The large total proton-proton cross-section at 14 TeV (100 mb) implies an observed event rate of about 10^9 events per second. The online event selection process (trigger) must reduce the huge rate to about 100 events per second for storage, reconstruction, and subsequent analysis. The short time between bunch crossings (25 ns) requires a fast readout and trigger system. Additionally, as mentioned in the previous section, to distinguish the hard interaction under study from the large number of soft pileup events requires high-granularity detectors with good time resolution to lower the particle occupancy. Finally, The large flux of particles from the interaction point also leads to high radiation levels, requiring radiation-hard detectors and front-end electronics.

To meet the goals of the LHC physics program, the CMS detector needs to have excellent muon identification and momentum resolution. It is required to efficiently identify photons and electrons with excellent energy resolution, efficiently tag charged-particles including τ leptons and b-quarks. The detector should also provide excellent missing transverse energy resolution with a hadron calorimeter that has large hermetic geometric coverage and fine lateral segmentation.

The design of the CMS detector, as shown in Figure 4.5 and detailed in the following sections, meets these requirements. Each detector subsystem is integral to the performance of CMS as a whole and is specialized to a particular class of particles: the silicon tracker measures the tracks of charged particles, the electromagnetic calorimeter measures the energy of electrons and photons, the hadron calorimeter measures the energy of charged and neutral hadrons, and the muon detector identifies and measures the momentum of muons.

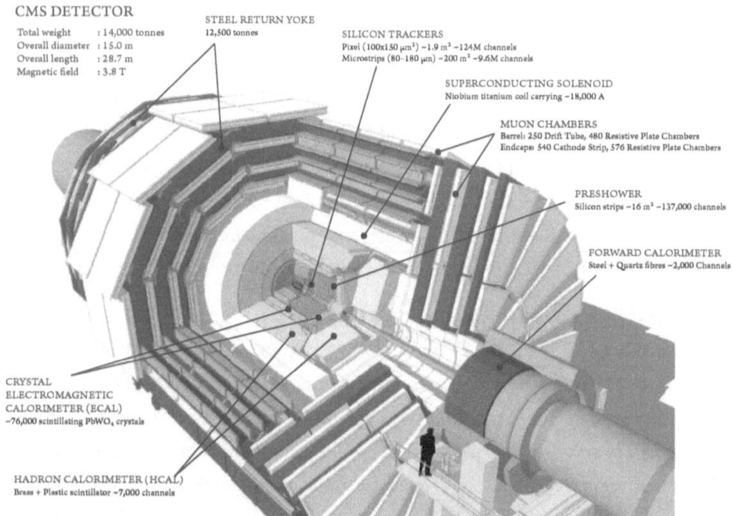

Figure 4.5: Cutaway diagram of CMS detector in Run 2 after Phase 1 Upgrade of the pixel detector at the end of 2016 [51]. The upgraded pixel detector is designed to cope with higher luminosity and has better tracker performance and lower mass [52].

The CMS detector is 21.6-m long and has a diameter of 14.6 m. It has a total weight of 12500 tons. At the heart of CMS sits a 13-m-long, 6-m-inner-diameter, 3.8-T superconducting solenoid providing a large bending power (12 Tm) before the muon bending angle is measured by the muon system. The bore of the magnet coil accommodates an all-silicon pixel and strip tracker, a lead-tungstate scintillating crystal electromagnetic calorimeter, and a brass-scintillator sampling hadron calorimeters. Outside of the solenoid, the return field of the solenoid is large enough to saturate 1.5 m of iron, allowing four muon stations to be integrated just outside of the solenoid to ensure robustness and full geometric coverage.

The coordinate system adopted by CMS uses a right-handed Cartesian coordinate system, as shown in Figure 4.6, where the origin is the nominal collision point, the y-axis pointing vertically upward, and the x-axis pointing radially inward toward the center of the LHC. Thus, the z-axis points along the beam direction toward the Jura mountains from LHC Point 5. The azimuthal angle ϕ is measured from the x-axis in the $x - y$ plane and the radial coordinate in this plane is denoted by r. The polar angle θ is measured from the z-axis. Pseudorapidity is defined as $\eta = - \ln \left[\tan \left(\frac{\theta}{2} \right) \right]$

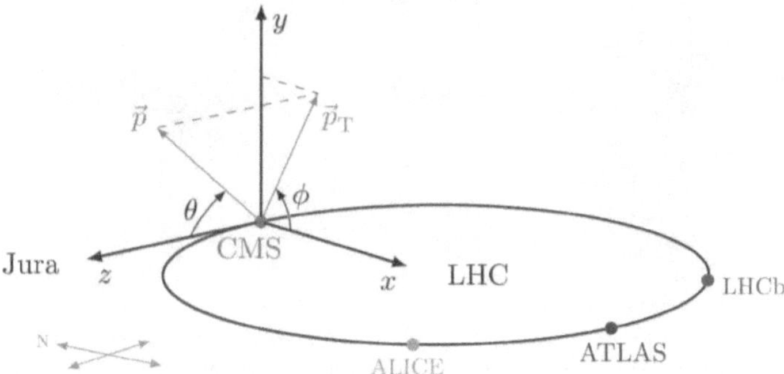

Figure 4.6: The coordinate system at the CMS detector [53].

The rest of this section is organized as follows: The superconducting magnet it introduced in Section 4.2.1, each of the sub-detectors of CMS are described in more detail in Section 4.2.2 to 4.2.5, the trigger and readout system is discussed in Section 4.2.6, and the event reconstruction techniques are detailed in Section 4.2.7.

4.2.1 Superconducting Magnet

In the CMS detector, the magnetic field is provided by a a wide-aperture super-conducting solenoid coil enclosed in a 10 kilo-ton steel flux-return yoke, which contributes to 70% of the total mass of the CMS detector. The solenoid has an inner bore of 6 m in diameter, a length of 12.5 m, a thickness of 0.313 m, and is operated with a direct current of 18.164 kA, creating a central magnetic flux density of 3.81 T. It is the largest and most powerful superconducting solenoid ever built. The solenoid has a total of 220 ton of cold mass and is composed of four layers of windings made from stabilized reinforced niobiumtitanium conductor. Outside of the solenoid, the flux is returned through the steel return yoke that also serve as the absorber plates of the muon detection system. The return yoke is composed of five dodecagonal three-layered barrel wheels and three end-cap disks at each end, that are made of steel plates up to 620 mm thick. Figure 4.7 shows a map of the magnetic field in CMS, which is required for the accurate simulation and reconstruction of physics events in the CMS detector.

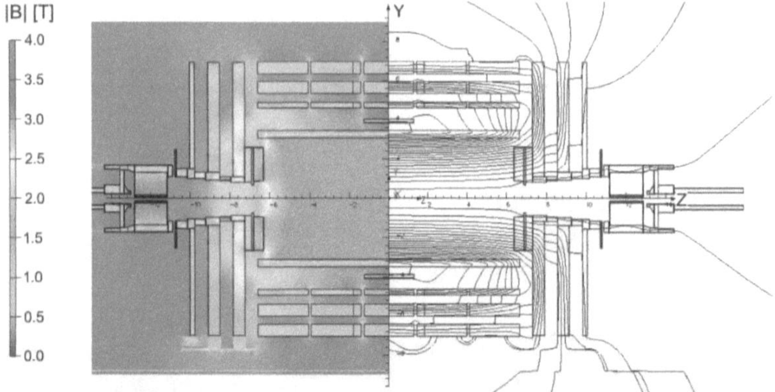

Figure 4.7: Value of $|B|$ (left) and field lines (right) predicted on a longitudinal section of the CMS detector, for the underground model at a central magnetic flux density of 3.8 T [54]. The modeled magnetic field map has been confirmed with measurements with high precision Hall probes and Teslameters [55, 56] and with cosmic muon [54].

4.2.2 Tracker

The CMS inner tracker is designed to precisely and efficiently measure the trajectory and momentum of charged particles and to precisely reconstruct secondary displaced vertices that are important for jet tagging and identifying long-lived particles. Due to the large number of pileup and short bunch crossing, a detector with high granularity, fast response, and radiation hardness for an expected lifetime of 10 years is required. Additionally, the amount of material in the tracker volume, including active detector elements, on-detector electronics and cooling materials, needs to be minimized to limit multiple scattering, bremsstrahlung, photon conversion, and nuclear interactions. These stringent requirements led to a tracker design composed of an inner pixel detector and an outer barrel detector, entirely based on silicon detector technology. The sketch of the CMS Phase-1 tracking system is shown in Figure 4.8.

The pixel detector is the part of the tracking system closest to the interaction point. The pixel detector is made of n-in-n sensors, with strongly n-doped (n+) pixelated implants on an n-doped silicon bulk and a p-doped back side. This design helps cope with the high radiation environment in the pixel and increases the charge sharing between neighboring cells to improve the spatial resolution. The cell size of the pixel

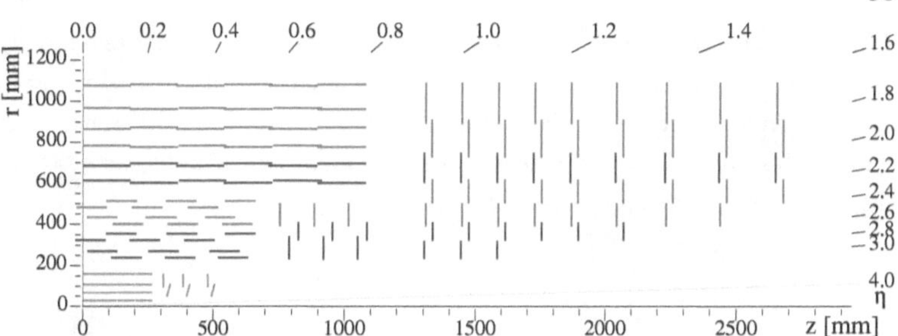

Figure 4.8: Sketch of one quarter of the Phase-1 CMS tracking system in r-z view. The pixel detector is shown in green, while single-sided and double-sided strip modules are depicted as red and blue segments, respectively [57].

detector is $100 \times 150 \, \mu\mathrm{m}^2$ in $r\phi \times z$ with a detector thickness of 285 mm [58]. When a charged particle passes through the pixel sensors, it usually generates charges in two or more pixels. Therefore reading out the analog signal allows for interpolation of amplitude to improve the spatial resolution to about $10 \, \mu\mathrm{m}$ ($20 \, \mu\mathrm{m}$) in $r\phi$ (z) direction.

The pixel detector covers a pseudorapidity range $-2.5 < \eta < 2.5$. It consists of three barrel layers (BPix) and two endcap disks (FPix). The 53-cm-long BPix layers are located at radii of 4.4, 7.3, and 10.2 cm. The FPix disks extending from 6 to 15 cm in radius, is placed on each side at z=±34.5 and z=±46.5 cm. BPix (FPix) contains 48 million (18 million) pixels covering a total area of 0.78 (0.28) m^2.

During February-March 2017, a new pixel detector was installed, as part of the CMS Phase-1 upgrade [52]. The new pixel detector uses the same sensor but an optimized layout design and improved readout electronics, in order to cope efficiently with the higher luminosity (above the design value) produced by the LHC. During Long Shutdown 1 (2013-2015), a new beam pipe with a smaller radius of 23 mm replaced the original 30-mm radius beam pipe. The smaller beam pipe allowed the innermost layer of the CMS Phase-1 pixel detector to be placed closer to the interaction point. The CMS Phase-1 pixel detector consists of four concentric barrel layers at radii of 2.9, 6.8, 10.9, and 16 cm, and three disks on each end at distances of 29.1, 39.6, and 51.6 cm from the center of the detector. The layout of the CMS Phase-1 pixel detector is compared to the one of the original pixel detector in Figure 4.9. The total silicon area of the CMS Phase-1 pixel detector increased from the original 1.1^2 to 1.9 m^2. Despite the additional layer in BPix, the material budget of the new

Phase-1 pixel detector in the central region is almost unchanged and significantly reduced in the forward region at $|\eta| > 1$. This improvement achieved by using advanced carbon-fiber materials for the mechanical structure, adopting the use of a lower mass, two-phase CO_2 cooling system, and moving the electronic boards to higher pseudorapidity regions outside of the tracking acceptance.

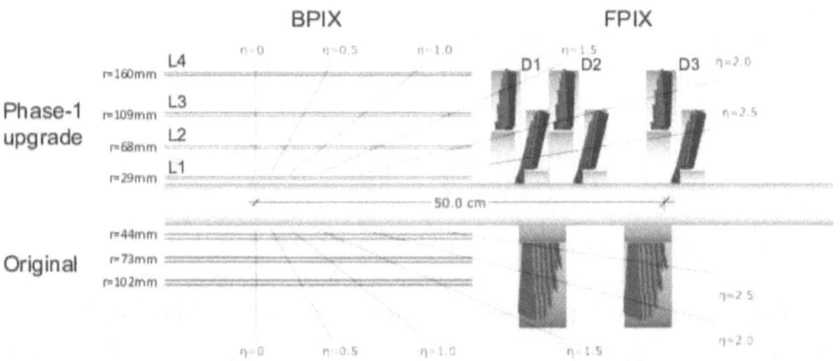

Figure 4.9: Layout of the CMS Phase-1 pixel detector compared to the original detector layout in longitudinal view [52].

The strip detector covers the radial region between 20 cm and 116 cm and is made of single-sided and double-sided p-on-n type silicon micro-strip sensors with thickness 320-500 μm and pitch sizes 80-183 μm. The strip detector is composed of 10 barrel layers and 3+9 endcap disks with a total of 10 million strips. About half of the modules are double-sided modules, made of two back-to-back single-sided modules with a relative rotation of 100 mrad. This allows an additional determination of the ionization in the z coordinate in the barrel modules, and in the r coordinate in the disks.

The material budget of the CMS tracker is shown in units of radiation and nuclear interaction lengths in Figure 4.10. Due to the materials in the tracker volume, a fraction of electrons and photon begin showering already in the tracker, which implies the need for a global strategy to event reconstruction, as detailed in Section 4.2.7.

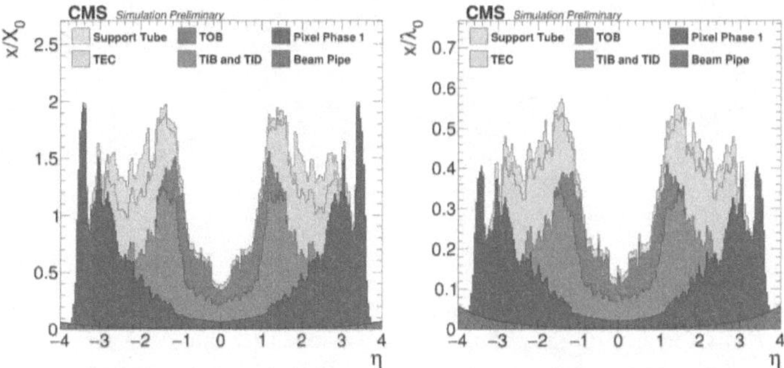

Figure 4.10: Material budget in number radiation lengths (x/X_0, left) of and hadronic interaction length (x/λ_0, right) as a function of the pseudorapidity η for Phase 1 tracker. The contribution of the support tube (light gray), the beam pipe (dark gray), and sub-detectors: TOB (red), Pixel Phase 1 (blue), TEC (yellow) and TID+TIB (magenta) are stacked [59].

Given the tracker's $O(10)\,\mu$m resolution and the bending power of the magnet, the CMS tracker can measure the transverse momentum of tracks with percent precision and the transverse impact parameter with 100-400 μm precision, as shown in Figure 4.11.

4.2.3 Electromagnetic Calorimeter

The CMS electromagnetic calorimeter (ECAL) is a hermetic homogeneous calorimeter made of lead tungstate ($PbWO_4$) crystals, segmented only in $\eta - \phi$, placed right outside of the tracker. One of the driving criteria in the ECAL design is its capability to detect the decay to two photons of the Higgs boson, which led to the use of a homogenous crystal calorimeter that has excellent energy resolution. $PbWO_4$ has high density (8.28 g/cm^3), short radiation length (0.89 cm), small Molière radius (2.2 cm), fast response (80% of light yield emitted in the first 25 ns), and high radiation tolerance, making it the ideal choice for a fine granularity and compact calorimeter. The scintillation light from the $PbWO_4$ crystals are detected by avalanche photodiodes (APDs) in the barrel and vacuum phototriodes (VPTs) that have higher radiation tolerance in the endcaps. These photodetectors are chosen due to their fast response, radiation tolerance, ability to operate in a 4 T magnetic field, and large gains to compensate for the small light yield from the crystals. Ad-

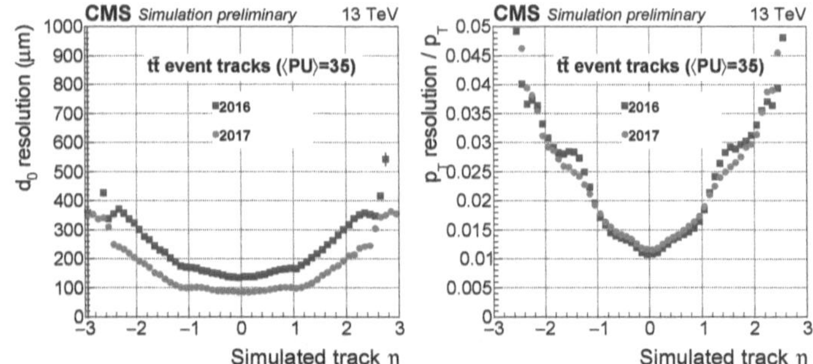

Figure 4.11: Track d_0 (transverse impact point; left) resolution and p_T resolution (right) as a function of the simulated track η for 2016 and 2017 detectors. The 2017 detector shows better performance than 2016 over all the η spectrum. The p_T resolution improves in $|\eta|$ 1.2-1.6, because the 4th pixel layer yields better precision on the track extrapolation to the strip tracker in the pixel barrel-forward transition region [60].

ditionally, a preshower detector is placed in front of each endcap to enhance photon identification again neutral pions.

The barrel part of the ECAL (EB) is composed of 61200 crystals, covering a pseudorapidity range $|\eta| < 1.479$ with the front faces of the crystals at a radius of 1.29 m. The crystals have a tapered shape, slightly varying with position in η. The crystal cross-section corresponds to about 0.0174×0.0174 in $\eta - \phi$ or $2.2 \times 2.2 \, \text{cm}^2$ at the front face and $2.6 \times 2.6 \, \text{cm}^2$ at the rear face of the crystal. The crystal length is 230 mm corresponding to 25.8 X_0. The endcaps of the ECAL (EE) are composed of 14848 crystals, covering a pseudorapidity range of $1.479 < |\eta| < 3.0$, with the front faces of the crystal at a z position of 315.4 cm. The crystals have a a front face cross section $2.86 \times 2.86 \, \text{cm}^2$, a rear face cross section $3 \times 3 \, \text{cm}^2$, and a length of 22 cm (24.7 X_0).

To identify photons and electrons against neutral pions and minimum ionizing particles, a preshower detector that has much finer granularity is placed in front of EE, covering $1.653 < |\eta| < 2.6$. The preshower detector is composed of two layers, each layer is composed of a lead absorber to initiate electromagnetic showers from incoming photons/electrons with silicon strip sensors (1.9 mm pitch size) behind to measure the deposited energy and the transverse shower profiles. The two layers of lead absorber have thickness of $2X_0$ (1.12 cm) and $1X_0$ (0.56 cm), respectively.

Thus about 95% of single incident photons start showering before the second sensor plane. The orientations of the strips in the two planes are orthogonal to measure the transverse shower profile. The total thickness of the preshower is 20 cm.

The typical relative ECAL energy resolution is measured to be approximately 2–5% during Run 2, as shown in Figure 4.12.

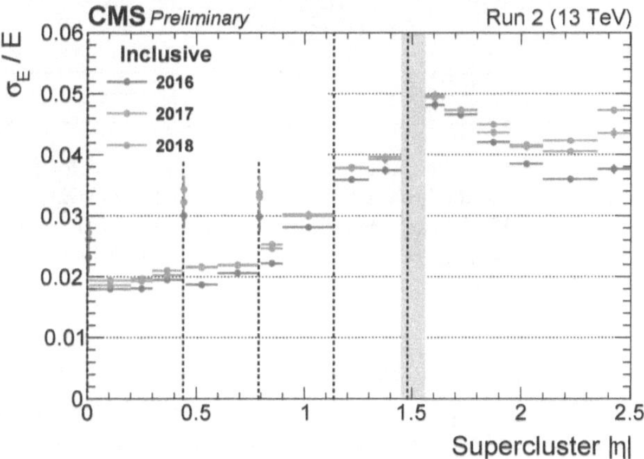

Figure 4.12: Relative electron (ECAL) energy resolution measured with electrons from $Z \to e^+e^-$ events as a function of the supercluster η [61]. The vertical lines indicate the boundaries between modules or boundary between EB and EE. A stable ECAL energy resolution is observed over the course of Run 2 despite the increased LHC luminosity and the aging of the detector.

4.2.4 Hadron Calorimeter

The hadron calorimeters (HCAL) in CMS that sits behind the ECAL is designed to measure hadronic jets and the calculation of missing transverse energy resulting from neutrinos or BSM particles. The CMS HCAL is divided into four subsystems: HCAL barrel (HB) covering $|\eta| < 1.3$, HCAL endcap (HE) for $1.3 < |\eta| < 3.0$, HCAL forward (HF) for $3.0 < |\eta| < 5.2$, and an outer hadronic calorimeter (HO) that is placed in the barrel region outside of the solenoid, but in front of the muon system for $|\eta| < 1.2$. The layout of the CMS detector showing the four subsystem of the HCAL is shown in Figure 4.13.

HB and HE, placed inside the solenoid, are sampling calorimeters with brass as absorber and plastic scintillators as the active layers. The sampling fraction is about

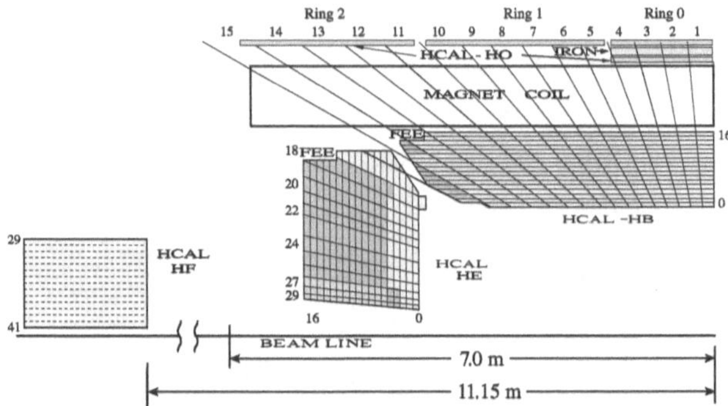

Figure 4.13: The quarter slice layout of the CMS HCAL detector [62]. "FEE" indicates the locations of the Front End Electronics for HB and HE. The signals of the tower segments with the same color are added optically, to provide the HCAL "longitudinal" segmentation. HB, HE and HF are built of 36 identical azimuthal wedges ($\Delta\phi = 20°$).

7%. The brass being used is called C26000/cartridge brass, composed of 70% copper and 30% zinc. It is non-magnetic and has a high density of 8.53 g/cm^3, radiation length of 1.49 cm, and nuclear interaction length of 16.42 cm. The HCAL barrel is radially restricted between the outer extent of the ECAL ($R = 1.77$ m) and the inner extent of the magnet coil ($R = 2.95$ m), constraining the total amount of material that can be put in to absorb and contain hadronic showers. Due to the space constraint in the solenoid, HB thickness is limited to 5.8 hadronic interaction lengths at $|\eta| = 0$ and increases to 10 interaction lengths at $|\eta| = 1.2$. Therefore, an outer hadron calorimeter (HO) is placed outside the solenoid, in front of the barrel muon system to complement the barrel calorimeter.

HO is a sampling calorimeter made of iron as the absorber and plastic as the active layers. The HO utilizes the cryostat and solenoidal coil as additional absorber (1.4 interaction length at $\eta = 0$) to identify and measure the late starting showers. Since the central region has the smallest value of interaction length from HB, two layers of HO scintillators were placed on either side of a 19.5 cm thick piece of iron at radial distances of 3820 mm and 4070 mm, respectively. All other rings have a single HO layer at a radial distance of 4070 mm. The total depth of the calorimeter system is thus extended to a minimum of 11.8 interaction lengths except at the barrel-endcap boundary region that has only about 8 interaction lengths. Due to limited space, the scintillator tiles in HB, HE, and HO are readout with embedded wavelength-shifting

(WLS) fibers and channeled to hybrid photodiodes (HPD) located a few meters away. The HPDs can provide high gain and operate in high axial magnetic fields.

Beyond $|\eta| = 3$, the HF placed at ± 11.2 m from the interaction point extends the pseudorapidity to $|\eta| = 5.2$ using a Cherenkov-based, radiation-hard technology. HF uses scintillating quartz fibers as the active material and steel as the absorber. The Cherenkov light produced by the charged shower particles in the quartz fibers is detected by photomultipliers. To separate the showers generated by electrons or photons that tend to produce closer to the front face of the HF from hadronic showers, two different lengths of quartz fibers are used. The long fibers (165 cm ≈ 10 interaction lengths) measure the total signal coming from the full material length, whereas the short fibers measure the energy deposition after 22 cm of steel. The forward calorimeters ensure full geometric coverage for the measurement of the transverse energy in the event. HB, HE, and HO have a segmentation of about $\Delta \eta \times \Delta \phi = 0.087 \times 0.087$ and HF has a segmentation of 0.175×0.175.

4.2.5 Muon System

Muons are an unmistakable signature of most of the physics LHC is designed to explore, especially the decay of the Higgs boson into ZZ*, which in turn decays into four leptons. Therefore, muon identification, momentum measurement, and triggering are central to the concept of CMS, the Compact Muon Solenoid. Good muon momentum resolution and trigger capability are enabled by the high-field solenoidal magnet and its flux-return yoke. The latter also serves as a hadron absorber for the identification of muons. The material thickness crossed by muons, as a function of pseudorapidity, is shown in Figure 4.14. The CMS muon system consists of about 25000 m^2 of detection planes, so the muon chambers had to be inexpensive, reliable, and robust. The CMS muon detector is composed of three types of gaseous particle detectors for muon identification: drift tubes (DT) in the barrel region ($|\eta| < 1.2$), cathode strip chambers (CSC) in the endcap region ($0.9 < \eta < 2.4$), and a complementary dedicated trigger system consisting of resistive plate chambers (RPC) in both barrel and endcap regions, as shown in Figure 4.15. The muon detectors play critical roles in the search described in Chapter 5.

In the barrel region, where the neutron-induced background is small, the muon rate is low, and the magnetic field is mostly uniform with strength below 0.4 T in between the yoke segments, drift chambers with standard rectangular drift cells are used. The

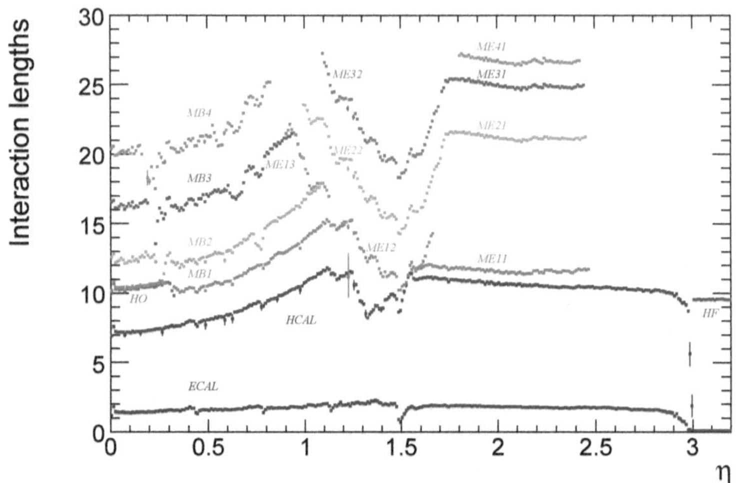

Figure 4.14: Material thickness in interaction lengths after the ECAL, HCAL, and at the depth of each muon station as a function of pseudorapidity. The thickness of the forward calorimeter (HF) remains approximately constant over the range $3 < |\eta| < 5$ (not shown) [63].

DT chambers are organized into four concentric cylindrical layers (stations) around the beamline and five wheels along the beamline axis (z). The four DT stations labeled MB1 to MB4 are located approximately 4, 5, 6, and 7 m away from the interaction point radially (r) and interleaved with the layers of the steel flux-return yoke. The basic element of the DT is the drift cell, that has a transverse size of 42 $\times$ 13 mm^2 with 50 μm diameter gold-plated stainless-steel anode wire at the center, as shown in Figure 4.16. As charged particles traverse the DT stations, they ionize the gas and produce charges that drift to the anode wire at the center of the DT cells. A signal pulse measured at the anode wire is recorded as a hit. The gas mixture (85%/15% of Ar/CO$_2$) provides good quenching properties. The electron average drift velocity of about 54.8 μm/ns [65]. The maximum drift time is almost 400 ns. The cell design makes use of four electrodes to shape the effective drift field: two on the side walls of the tube, and two above and below the wires on the ground planes between the layers. Four staggered layers of parallel cells form a superlayer (SL). The first three stations, each containing 12 layers of DT cells, are arranged in three groups of SLs, as shown in Figure 4.16. The innermost and outermost

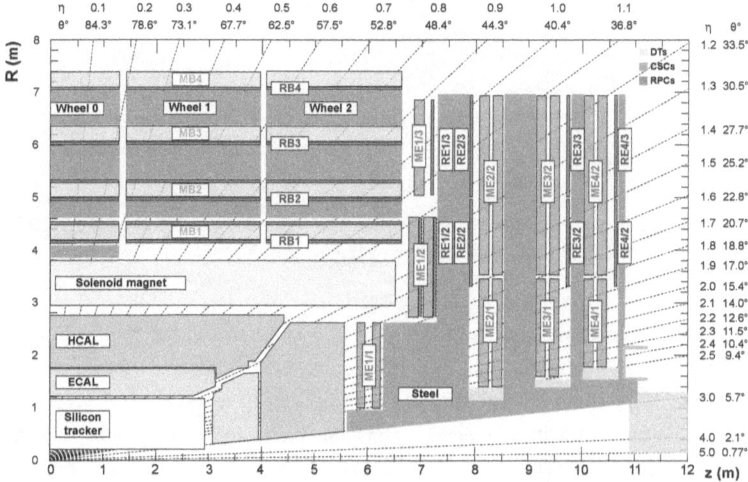

Figure 4.15: An rz cross section of a quadrant of the CMS muon detector [64]. Shown are the locations of the various muon stations and the steel disks (dark grey areas). The four drift tube (DT, in light orange) stations are labeled MB (muon barrel) and the cathode strip chambers (CSC, in green) are labeled ME (muon endcap). Resistive plate chambers (RPC, in blue) are in both the barrel and the endcaps of CMS, where they are labeled RB and RE, respectively.

SLs measure the hit coordinate in the $r - \phi$ plane, and the central SL measures the position in the z direction, along the beamline. The fourth station only contains two SLs measuring the hit position in the $r - \phi$ plane. The SLs are glued together to a thick honeycomb plate that provides mechanical stiffness and increases the lever arm in the bending plane. Individual hits have a resolution of about $530\,\mu$m, which is largely dominated by the uncertainty on the time of arrival of the charged particle, $\sigma = (25\,\text{ns}/12) \times v_{\text{drift}} = 395\,\mu$m, where $v_{\text{drift}} = 54.8\,\mu$m/ns [65]. The hit resolution can be improved to about $260\,\mu$m by incorporating information from hits in other layers to perform a refit of the segment that determines the final segment position and direction. The DT hits are central to the reconstruction of the main object, muon detector shower, of the search discussed in Chapter 5.

In the endcap regions, the muon rates and background levels are high and the magnetic field is large and non-uniform. Therefore, cathode strip chambers (CSC) that have fast response time, fine segmentation, and radiation resistance are used in

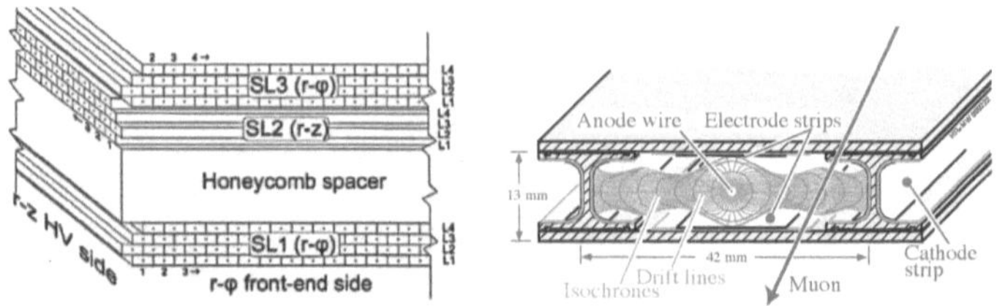

Figure 4.16: Schematic view of a DT chamber (left) and a section of a drift tube cell showing the drift lines and isochrones (right) are shown [64].

the endcaps between $|\eta|$ values of 0.9 and 2.4. The gas mixture of 50% CO2, 40% Ar, and 10% CF4 is used. The CSCs are organized in four stations in each endcap. The four CSC stations labeled ME1 to ME4 are located approximately 7.0, 8.0, 9.5, and 10.5 m away from the interaction point along the beamline axis in both endcaps, and are interleaved between steel absorbers. In the r direction, each station is composed of two or three rings, labeled as ME1/n-ME4/n, where integer n increases with the radial distance from the beam line, as shown in Figure 4.15. There are 270 CSC chambers in each endcap. The inner rings of stations 2, 3, and 4 are composed of 18 chambers subtending a ϕ angle of 20° and all other CSC chambers subtend 10° in ϕ. Each chamber is composed of six thin layers, each composed of an anode wire plane stretched between two planar copper cathodes, one continuous, the other segmented in strips to provide position measurement in two coordinates, as shown in Figure 4.17. The distance between anode planes is 2.54 cm, except for the ME1/1 chambers, for which it is 2.2 cm. The cathode strips run along the radial direction and provide precision measurement in the $r - \phi$ bending plane by exploring the shape of the charge distribution on three consecutive strips. There are 80 cathode strips per layer, each of which subtends a ϕ angle between 2.2 and 4.7 mrad, corresponding to a width of 3 to 16 mm, depending on the radius from the beamline and the location of the chamber. The anode wires are directly wired together in sets of 5 to 16 wires per readout channel, with widths from 16 to 51 mm, providing a coarse measurement in the radial direction. In addition, to reduce the rate in any one strip, the strip region of ME1/1 chambers are divided into two at $|\eta| = 2.1$, so that each region can trigger and be read out independently of the other. The innermost region is labelled "ME1/1a" and the outer "ME1/1b". Narrower strips with widths of 4.11-7.6 mm are used in both regions. ME1/1a consists of 48 strips per chamber and ME1/1b consists of 64 strips per chamber. Charged particles traversing the chambers ionize the gas. The resulting electrons are accelerated towards the anode wires producing an avalanche, while the positive ions travel to the opposite end and induce signals in the cathode strips. By combining the information from signals on the anode wires and the cathode strips of each layer, the space resolution of each hit is $130\,\mu$m in ME1/1, $270\,\mu$m in ME1/2, and 400–$600\,\mu$m in the other channels [66]. The time resolution of each hit is 5 ns [66, 67]. The CSC hits are central to the reconstruction of the main object, muon detector shower, of the search discussed in Chapter 5.

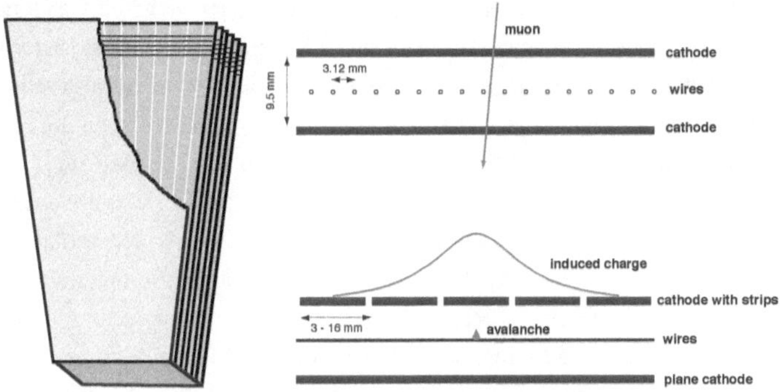

Figure 4.17: Left: A cut-away diagram of a CSC showing the six layers and the orientation of the wires and strips. Right: A cross-sectional view of the gas gap in a CSC showing the anode wires and cathode planes and an illustration of the gas ionization avalanche and induced charge distribution on the cathode strips [64].

In addition to the DTs and CSCs, the CMS muon system includes a complementary, dedicated triggering detector system which consists of RPCs that provide excellent timing resolution to reinforce the measurement of the correct beam crossing time. The RPCs are double-gap chambers, where each gap consists of two 2-mm thick resistive Bakelite plates separated by a 2-mm thick gas gap. The RPCs are operated with a non-flammable gas mixture that consists of 95.2% Freon ($C_2H_2F_4$), 4.5% isobutane ($i-C_4H_{10}$), and 0.3% sulphur hexafluoride (SF_6), where the highly electronegative SF_6 helps prevent breakdowns in the gas. The RPCs are located in both barrel and endcap regions with $|\eta| < 1.9$. The RPCs are arranged in stations following a sequence similar to the DTs and CSCs. There are four stations in the RPC barrel (RB) and four stations in the RPC endcap (RE). The two inner barrel stations RB1 and RB2 are instrumented with two layers of RPCs facing the inner and outer sides of the DT chambers, while all other stations are composed of one layer of RPC, as shown in Figure 4.15. The RPC hit has a spatial resolution of about 1 cm and a time resolution of 2 ns [68].

The geometry of CMS strongly influences the performance of the muon system. The curvature of the muon trajectory reverses in the muon system due to the change in the direction of the magnetic field from the return yoke. Therefore, the first muon detector stations in both the barrel and endcap regions (ME1, MB1) are critical, since they provide the largest sagitta and thus the most important contribution to the measurement of the momentum of high p_T (more than a few hundred GeV) muons. Muon tracks can be reconstructed by using hits in the muon detectors alone, resulting in muon candidates called *standalone muons*. Alternatively, the reconstruction can combine hits in the muon detectors with those in the inner tracker, resulting in muon candidates called *global muons*. The muon system can also be used simply to tag tracks extrapolated from the inner tracker, which are called *tracker muons*. For muons with momenta below $\approx 200\,\text{GeV}$, tracker muons have better resolution than global muons. As the p_T value increases, the additional hits in the muon system gradually improve the overall resolution. Global muons exploit the full bending of the CMS solenoid and return yoke to achieve the ultimate performance in the TeV region. The momentum resolution can be extracted from the distribution of the relative residual in q/p_T, denoted as $R(q/p_T)$, where q and p_T are the charge and momentum of the muon, respectively. Figure 4.18 shows the RMS of $R(q/p_T)$ as a function of p_T for cosmic ray muons recorded in 2015 for fits using only the inner tracker and for fits that include the muon system using the Tune-P algorithm. The uncertainty in the last bins is dominated by the small number of cosmic rays collected in 2015 (66 events with $p_T > 500$ GeV). The improvement in resolution from exploiting the muon detector information in the momentum assignment for high p_T muon is clearly visible.

4.2.6 Trigger and Data Acquisition

4.2.6.1 Trigger System

The LHC provides pp collisions at a high crossing rate of 40 MHz. Given that 1 MB disk space is needed to store the raw digitized signals from all CMS subdetectors in one bunch crossing, it is impossible to save and process the events from all bunch crossings, which would require storing and transferring tens of TB of data per second. Additionally, only a small fraction of the collisions contain events of interest to the CMS physics program, so a trigger system is deployed to apply

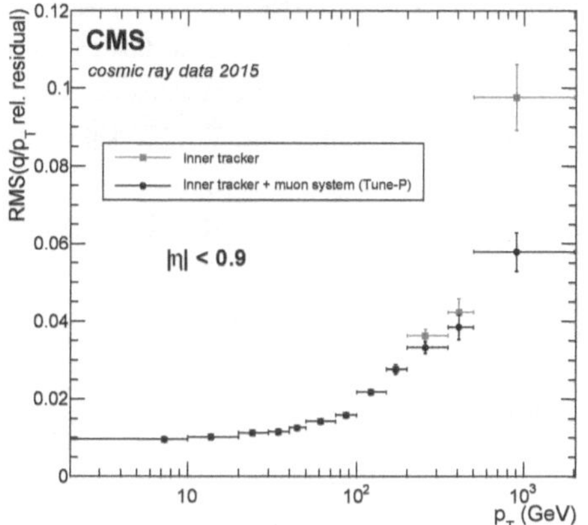

Figure 4.18: The RMS of $R(q/p_T)$ as a function of p_T for cosmic rays recorded in 2015, using the inner tracker fit only (red squares) and including the muon system using the Tune-P algorithm (black circles). The vertical error bars represent the statistical uncertainties of the RMS [67].

online selection of the events before they are stored and processed for later offline analysis. The CMS trigger system consists of two sequential but independent levels. The Level-1 (L1) trigger, based on custom electronics, reduces the event rate from 40 MHz to 100 kHz with a 3.8 μs latency, and the High-Level Trigger (HLT), implemented in software running on computer farm based on commercial CPUs, further reduces the rate to 1.5 kHz for offline processing and storage.

The L1 system receives information at a rate of 40 MHz from the calorimeters and muon detectors with coarse granularity and precision to select interesting collision events. Events are selected based on the presence of energy deposits compatible with physics objects such as photons, electrons, muons, jets, hadronic τ, scalar sum of transverse energy (H_T), and the energy corresponding to the vector sum of the transverse missing momentum (p_T^{miss}). Any event that satisfies the conditions of at least one seed (predetermined criteria) in the trigger menu is accepted for further processing in the trigger chain, which initiates a readout of the complete detector information from the data acquisition system, and the data are sent to the HLT. Figure 4.19 shows the fraction of the maximum Level-1 trigger rate allocated to different object seeds. The broad range of menu algorithms reflects the wide variety of research interests of the CMS experiment. The Level-1 menu also evolves with shifting CMS physics priorities, new physics ideas, and changes in beam or detector performance. As detailed in Chapter 7, a new dedicated L1 seed to search for long-lived particles decaying in the CMS muon detectors are added to the menu in Run 3 to open up a new phase space that were not accessible in Run 2.

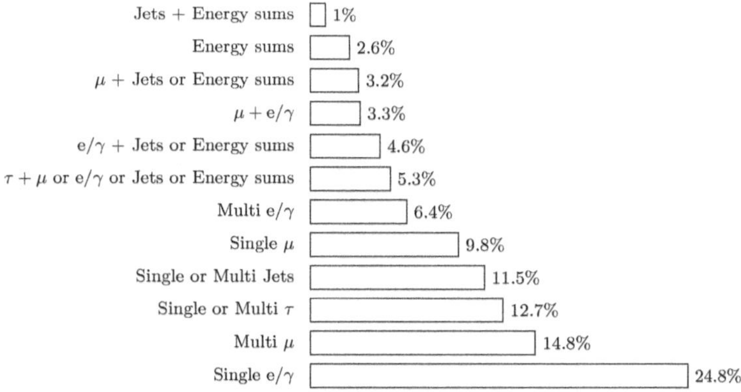

Figure 4.19: Fractions of the 100 kHz rate allocation for L1 triggers in a typical CMS physics menu during Run 2. [69].

The L1 trigger in Run 2 uses Advanced Mezzanine Cards (AMC) based on MicroTCA technology and multi-Gb/s serial optical links for data transfer between modules. All processor cards use a Xilinx Virtex-7 Field Programmable Gate Array (FPGA), allowing many firmware and control software components to be reused by several systems, reducing the workload for development and maintenance. The L1 trigger is composed of the calorimeter trigger and muon trigger. The architecture of the L1 trigger system during Run 2 is shown in Figure 4.20.

The calorimeter trigger consists of two layers. Layer-1 receives, calibrates, and sorts the trigger primitives (TPs), that are local energy deposits sent to the trigger from the calorimeters. Each TP accesses the energy deposits with a $\Delta\eta \times \Delta\phi$ granularity of 0.087 × 0.087 radians in most of the calorimeter acceptance (slightly coarser granularity is used at high $|\eta|$), which corresponds to 5 × 5 ECAL crystals and the one HCAL tower directly behind them in the barrel region. The calibrated TPs are then sent to Layer-2 with a time-multiplexed algorithm [70], where physics objects such as electrons, τ leptons, jets, and energy sums, are reconstructed and

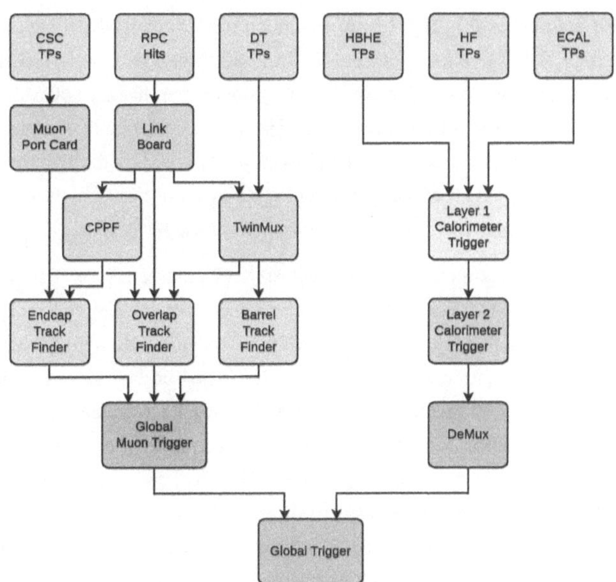

Figure 4.20: Diagram of the CMS Level-1 trigger system during Run 2 [69].

calibrated. The physics objects are then sent to a demultiplexer (DeMux) board that then reorders, reserializes, and formats the events for the global trigger (μGT) processing.

Similarly, the muon trigger also starts with TPs. TPs in the muon detectors (CSCs, RPCs, DTs) provide spatial coordinates, timing, and quality information from detector hits. The TPs from all available sub-detectors are used to reconstruct tracks in three track finders covering distinct pseudorapidity regions to improve the muon reconstruction efficiency and resolution while reducing the misidentification rate. The muon track finders build muon track candidates, assign a quality to each, and measure the charge and p_T of each candidate from the bending in the fringe field of the magnet yoke. The barrel muon track finder (BMTF) reconstructs muons in $|\eta| < 0.83$ using inputs from a board called TwinMux [71] that merges DT and RPC TPs from the same station into *superprimitives* that combine the better spatial resolution of the DT and the more precise timing from the RPC. The endcap muon track finder (EMTF) reconstructs endcap muons in $|\eta| > 1.24$ using CSC TPs collected and sorted by the Muon Port Cards and RPC hits from CPPF (concentrator preprocessor and fan-out) card [72]. The overlap muon track finder (OMTF) reconstructs muon in $0.83 < |\eta| < 1.24$ by using TPs from all three muon subsystems. Each track finder transmits up to 36 muons to the global muon trigger (μGMT), which resolves duplicates from different boards, and sends the data for a maximum of eight muons of highest rank (ranked by a combination of p_T and a quality value) to the μGT. The μGT collects all of the muons and calorimeter objects and executes every algorithm in the trigger menu in parallel for the final trigger decision.

The HLT is based on commercial CPUs that run on Scientific Linux. The HLT farm consists of about 30,000 CPUs in Run 2 and selects events accepted by L1 using the full precision of the data from the detector based on offline-quality reconstruction algorithm. The processing time of the HLT decision is required to be about 300 ms, constrained by the computing power of the HLT farm and the L1 output rate. The HLT selection is made with a trigger menu with about 400 HLT paths, targeting a broad range of physics signatures. The HLT paths are constructed in a modular fashion, consisting of sequences of reconstruction and filter modules that are arranged and executed in increasing complexity. If a particular event is rejected by a filter in an HLT path, the subsequent modules in the same HLT path are not run. The missing transverse momentum trigger paths are being used and discussed in more detail in Chapter 5. Additionally, two new HLT trigger paths seeding on the the new L1 seed to search for long-lived particles decaying in the CMS muon detectors were commissioned in Run3 and will be discussed further in Chapter 7.

The events files passing HLT paths are sorted into *streams* that consists of a set of HLT paths and a well-defined event content. Primary datasets (PDs) are created and associated with specific streams for efficient data handling. The HLT paths targeting similar physics processes are grouped into common PDs. The PDs are defined to keep the total event rate balanced and within the limits imposed by the data offline processing. Events can in principle end up in more than one PD due to different trigger selections, but significant effort is made to minimize the event overlap. Additionally, the HLT also contains specific paths and data streams to gather information for detector calibrations and to perform online monitoring of the data quality during data taking.

4.2.6.2 Data Acquisition System

The data acquisition (DAQ) system of the CMS experiment performs the readout and assembly of events after they are accepted by the L1 trigger. Assembled events are made available to the HLT which selects for interesting events for offline storage and analysis. The DAQ system is designed to handle a maximum input rate of 100 kHz and an aggregated throughput of 100 GB/s. L1-accepted events are assembled through a 2-stage event-building system that are then fed into the HLT farm, which reduces the event rate to ~1 kHz. The HLT-accepted events are then written to a temporary disk buffer before being transferred to the computing center (Tier 0) at CERN for offline processing.

The 2-stage event builder assembles event fragments from the detector front-end located underground into one super-fragment which is then fed into one of the readout slices to the surface where the complete event is built, as shown in Figure 4.21. The digitized output signals from all of the sub-detectors are continuously stored in the sub-detector 40-MHz front-end pipelined buffers. In addition, the calorimeters and muon detectors also send their output data to the L1 trigger, which gives an L1 decision about $4\mu s$ later to the detector front-ends via the Timing, Trigger, and Control (TTC) system. Upon arrival of the L1 acceptance decision, the data in the detector front-end buffers are extracted and pushed into the Front-End Drivers (FEDs). Data fragments from the FEDs are then transported from the electronics room in the Underground Service Cavern (USC) that are about 90 m underground to the surface DAQ building (SCX), where the FED builders are located. The FED builder assembles the data fragments from the FEDs into super-fragments and send the super-fragments to the buffer in readout units (RU) running on commodity PCs. The RU-builder then build the full events out of super-fragments, perform physics selections, and forward the selected events to mass storage.

4.2.6.3 Data Format

Data selected by the HLT farm are sent to the Tier-0 (T0) CERN computing center. A prompt reconstruction on about 10% of the events is performed to monitor the offline data quality and calculate fast-turnaround detector alignment and calibrations. Once the prompt calibrations are calculated, the prompt reconstruction of the entire dataset is performed within 48 hours of data-taking to avoid overflowing the data storage buffers. The data are then distributed to the seven Tier-1 computing sites around world that maintain a second copy of the RAW data on tape and provide CPUs for future reprocessing. The RAW data are processed and saved in different data formats where successive degrees of processing refine the data, apply calibrations and create higher-level physics objects. The RECO dataset is the largest dataset (4 MB/event)

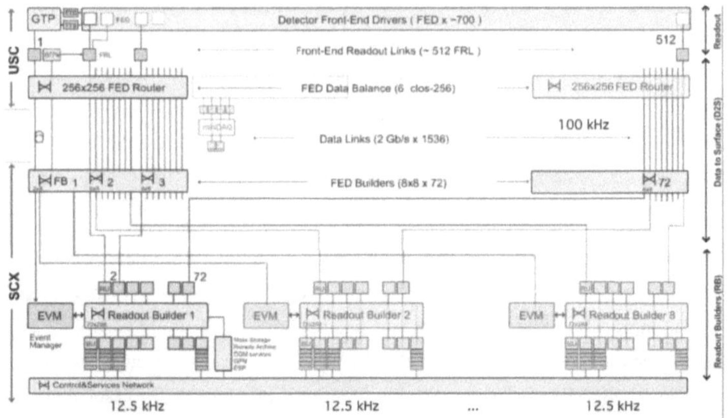

Figure 4.21: The 2-stage event builder assembles event fragments from typically 8 front-ends located underground (USC) into one super-fragment which is then fed into one of the 8 independent readout slices on the surface (SCX) where the complete event is built [73].

after the reconstruction step that collects all reconstructed objects from all stages of reconstruction, including reconstructed hits (rechits), clusters, and segments in each subdetector and high-level reconstructed objects like jets, muon, electrons, and etc. Due to the large size of the RECO datasets and the small number of analyses that use these datasets, most of the RECO datasets are stored only on tape. The AOD dataset (0.4 MB/event) is derived from the RECO dataset, containing all high-level reconstructed objects and a small quantity of rechit information, and is readily available in all Tier-1 computing centers. The MINIAOD format (40 kB/event) was created at the beginning of Run 2 that provides sufficient event information to cover 95% of physics analyses performed by CMS during Run 2 [74]. By the end of Run 2, the NANOAOD format (1-2 kB/event) that stores only high-level physics objects was commissioned to cover the needs of 50-70% of physics analyses [74]. The analysis described in Chapter 5 needs to access low-level rechits information from the muon detectors, so it uses the RECO dataset.

4.2.7 Event Reconstruction

Once the detector records an event, the ability to reconstruct and identify each individual particle in an event is crucial to perform physics analysis. CMS uses a holistic and global event reconstruction technique called particle-flow event reconstruction [75] that optimally combines all subdetector information to reconstruct particle properties. This reconstruction technique makes use of the fact that different particles exhibit different behaviors as they traverse through the cylindrical detection layers nested around the beam axis. A sketch of the different particle interactions in a transverse slice of the CMS detector is shown in Figure 4.22.

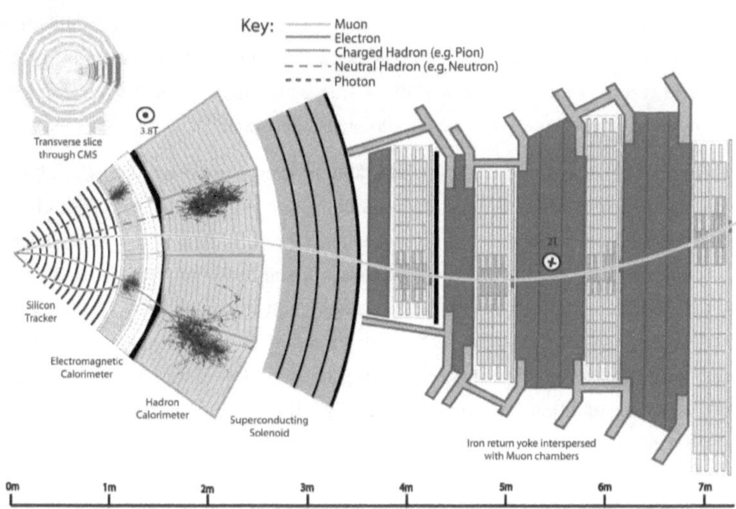

Figure 4.22: A sketch of the specific particle interactions in a transverse slice of the CMS detector, from the beam interaction region to the muon detector. The muon and the charged pion are positively charged, and the electron is negatively charged [75].

The PF algorithm starts with the local reconstruction of basic PF elements within each subdetector: charged-particle trajectories (tracks) in the inner tracker, calorimeter clusters in the ECAL and HCAL, and tracks in the muon detectors. Track reconstruction is a process of combining hits in different layers of the tracker or muon detectors to obtain charged-particle trajectory and measure the momentum and direction of the charged particles. Tracks in the central tracker and in the muon detectors are reconstructed independently using the track finder algorithm based on a combinatorial Kalman Filter [76]. Tracker tracks are reconstructed in an iterative process from a list of seeds that are pairs or triplets of hits passing some kinematic selections. From each seed, the track finder extrpolates the seed hits to the outer layers to find compatible hits with the Kalman Filter. When all the layers are taken into account, a quality selection based on the goodness-of-fit and the number of missing hits is applied to improve the track purity and reject fake tracks. Tracks in the muon detectors (standalone muon) are reconstructed using seeds from DT or CSC segments, straight-line tracks built from rechits in each layer within a DT or CSC chamber, and applies the Kalman Filter. A global muon track is reconstructed if a standalone muon track is compatible to a tracker track propagated to the muon detectors. For $p_T > 200\,\text{GeV}$, the global-muon fit improves the momentum resolution with respect to the tracker-only fit. A tracker muon is reconstructed if at least one muon segment matches to an extrapolated inner track.

Calorimeter clusters are reconstructed separately in each subdetector: ECAL barrel and endcaps, HCAL barrel and endcaps, and the two preshower layers. The cluster reconstruction starts from identifying a cluster seed, which is a cell with local maximum energy and has an energy above a given seed threshold. Topological clusters are then grown from the seeds by aggregating cells with at least a corner in common with a cell already in the cluster and with an energy above twice the noise level. In the ECAL endcaps, because the noise level increases as a function of θ, seeds are additionally required to satisfy a threshold requirement on E_T. For topological clusters that contain multiple seeds, each seed is assumed to represent a unique energy cluster, but the energy deposited in non-seed cells must be shared between the various clusters within the topological cluster. An iterative procedure is used to converge on cluster energies and positions based on energy-weighted averages of fractional cell energies. When there is little or no bremsstrahlung, electron and photon showers deposit their energy in several crystals in the ECAL, with 97% of the incident energy contained in 5×5 crystals [77]. However, due to the large amount of tracker materials (0.4–2 radiation lengths), on average, 33% of the electron energy is radiated before it reaches the ECAL when the intervening material is minimal ($\eta \approx 0$), and about 86% of its energy is radiated when the intervening material is the largest ($|\eta| \approx 1.4$) [77]. Therefore, to account for the bremsstrahlung photons, superclusters are reconstructed by grouping ECAL clusters in a small window in η and an extended window in ϕ around the electron direction (to account for the azimuthal bending of the electron in the magnetic field).

Since a given particle is in general expected to give rise to several PF elements in different CMS subdetectors, the reconstruction of a particle therefore proceeds with a *link algorithm* that connects the PF elements from different subdetectors. A link can be a track-cluster link between the central tracker and calorimeter cluster, cluster-cluster link between ECAL and HCAL cluster, track-track link between central tracker tracks that share a common secondary vertex, or tracker track and muon detector track link. If multiple links are found, only the link with the smallest distance is kept, for example when multiple HCAL clusters are linked to the same ECAL cluster, only the link with the smallest distance is kept. To prevent the computing time of the link algorithm from growing quadratically with the number of particles, the pairs of elements considered are restricted to the nearest neighbors in the $\eta - \phi$ plane. The link algorithm then produces PF blocks of elements associated either by a direct link or by an indirect link through common elements.

Within each PF block, the identification and reconstruction sequence for each physics objects is performed and the corresponding PF elements are removed from the PF block in the following order: muons, electrons, isolated photons (converted or unconverted), and hadrons and non-isolated photons from jet fragmentation and hadronization. Muons are identified as tracks in the central tracker consistent with either a track (global muon) or several hits (tracker muon) in the muon system, and associated with calorimeter deposits compatible with the muon hypothesis. The muon momentum is assigned to be that of the inner tracker for low p_T muons ($p_T < 200\,\text{GeV}$). High p_T muons momentum are assigned based on the goodness-of-fit and momentum uncertainty information from the different track fits [67]. Once muons are identified, the PF elements that make up these muons are not used as building elements for other particles. In a given PF block, an electron candidate is seeded from a track with a link to ECAL energy clusters and a photon candidate is seeded from an ECAL supercluster with $E_T > 10\,\text{GeV}$ with no link to a track. Additionally, the energies from HCAL cells that are in proximity to the ECAL cluster position ($\Delta R < 0.15$) is required to be less then 10% of the supercluster energy. The energy and direction of the photons is taken to be that of the ECAL clusters, while the energy of electrons is determined from a combination of the track momentum and the corresponding ECAL cluster energy. Once muons, electrons, and isolated photons are identified, the remaining particles are hadrons from jet fragmentation and hadronization. The ECAL and HCAL clusters that are not linked to any track give rise to photons and neutral hadrons. Within the tracker acceptance ($|\eta| < 2.5$), all such ECAL clusters are identified as photons and all such HCAL clusters are identified as neutral hadrons. Beyond the tracker acceptance, however, charged and neutral hadrons cannot be distinguished and they leave in total 25% of the jet energy in the ECAL. Therefore, ECAL clusters linked to an HCAL cluster are assumed to arise from the same (charged- or neutral-) hadron shower, while ECAL clusters without such a link are classified as photons.

The presence of particles that do not interact with the detector material, for example neutrinos, is indirectly detected through missing transverse momentum (p_T^{miss}), which is defined as the negative vectorial sum of the transverse momenta of all particles:

$$p_T^{miss} = - \sum_{i=1}^{N_{particles}} \vec{p}_{T,i}. \tag{4.4}$$

Usually the jet energy corrections, as described in the following paragraphs, are propagated to the calculation of p_T^{miss} for all jets with $p_T > 10\,\text{GeV}$. The jet energy corrected p_T^{miss} is used in the analysis described in Chapter 5.

Finally, once all particles are reconstructed and identified, a post-processing step revisits the reconstructed event if an artificially large missing transverse momentum (p_T^{miss}) is found due to the presence of cosmic ray muon traversing CMS in coincidence with the LHC beam crossing, mis-reconstruction of muon momentum, or particle mis-identification that leads to the wrong momentum measurement.

More complex physics objects like jets and hadronically decaying τ leptons (τ_h) are reconstructed based on the PF objects. Jets are clustered from the PF objects using the infrared and collinear safe anti-k_T algorithm [78, 79] with a distance parameter of 0.4. The algorithm clusters either all particles reconstructed by the PF algorithm (PF jets), or the sum of the ECAL and HCAL energies deposited in the calorimeter towers (Calo jets). Particle-flow jets are used throughout this book, so hereafter if not mentioned specifically, jets refers to PF jets only. Jet momentum is determined as the vectorial sum of all particle momenta in the jet, and is found from simulation to be, on average, within 5 to 10% of the true momentum over the whole p_T spectrum and detector acceptance. Additional proton-proton interactions within the same or nearby bunch crossings (pileup) can contribute additional tracks and calorimetric energy depositions to the jet momentum. To mitigate this effect, charged particles identified to be originating from pileup vertices are discarded and an offset correction is applied to correct for remaining contributions. Jet energy corrections are derived from simulation to bring the measured response of jets to that of particle level jets on average. In situ measurements of the momentum balance in dijet, photon + jet,

Z + jet, and multijet events are used to account for any residual differences in the jet energy scale between data and simulation [80]. The jet energy resolution amounts typically to 15–20% at 30 GeV, 10% at 100 GeV, and 5% at 1 TeV [80]. Additional selection criteria are applied to each jet to remove jets potentially dominated by anomalous contributions from various subdetector components or reconstruction failures.

Hadronically decaying τ (τ_h) can have a number of possible final states with one or three charged pions and zero to three neutral pions. Therefore, τ_h are reconstructed from jets, using the hadrons-plus-strips algorithm (HPS) [81] that combines one or three tracks with energy deposits in the calorimeters, to identify the τ decay modes. Neutral pions are reconstructed as strips with dynamic size in η-ϕ from reconstructed electrons and photons, where the strip size varies as a function of the p_T of the electron or photon candidate. To distinguish τ_h decays from jets originating from the hadronization of quarks or gluons, and from electrons, or muons, the DEEPTAU algorithm is used [82]. The rate of a jet to be misidentified as τ_h by the DEEPTAU algorithm depends on the p_T and quark flavor of the jet. In simulated events from W boson production in association with jets it has been estimated to be 0.43% for a genuine τ_h identification efficiency of 70%. The misidentification rate for electrons (muons) is 2.60 (0.03)% for a genuine τ_h identification efficiency of 80 (>99)%.

Search for Neutral Long-lived Particles Decaying in the Muon Detectors at the CMS with Run 2 Data

5.1   Introduction

Many extensions of the standard model (SM) predict the existence of weakly coupled particles that have long proper lifetimes. These long-lived particles (LLPs) naturally arise in a broad range of models beyond the SM including supersymmetry (SUSY) [83–97], hidden valley scenarios [98–100], inelastic dark matter [101], and twin Higgs models [102–104]. A more comprehensive overview of models predicting LLPs can be found in Ref. [105, 106].

Similar to SM particles, LLPs maybe be electrically charged or neutral, or even colored in the case of long-lived gluinos. Charged and/or colored LLPs are particularly prevalent in supersymmetry models, while neutral LLPs tend to show up in models of dark matter, baryogenesis and certain non-SUSY solutions of the hierarchy problem.

As a probe to hidden sector dark matter, in this chapter, a search for neutral LLPs using the CMS muon detectors as a sampling calorimeter to identify particle showers produced by decays of LLPs is presented. The search is based on proton-proton (pp) collision data collected at a center-of-mass energy of 13 TeV during 2016–2018 at the LHC, corresponding to an integrated luminosity of $138\,\mathrm{fb}^{-1}$. The CMS muon detectors are composed of gaseous detector chambers interleaved with steel layers of the magnet flux-return yoke. Decays of LLPs in the muon detectors induce hadronic and electromagnetic showers, giving rise to a large multiplicity of hits in a localized detector region. The hadron calorimeter (HCAL), solenoid magnet, and steel flux-return yoke together provide 12–27 nuclear interaction lengths of shielding [46, 107], which, together with explicit vetoes on inner detector activity, strongly suppresses particle showers from jets that are not fully contained within the calorimeters' volume (punch-through). An LLP produced with a large Lorentz boost and decaying after it has traversed the calorimeter systems may produce large missing transverse momentum because its momentum is not properly measured or associated with a reconstructed particle. Therefore, the analyzed data are required to have a magnitude of the missing transverse momentum vector above 200 GeV.

Text and figures from this chapter are adapted with permission from CMS Collaboration. *Search for long-lived particles decaying in the CMS muon detectors in proton-proton collisions at $\sqrt{s} = 13$ TeV.*. Feb. 2024. arXiv: `2402.01898 [hep-ex]`.

The search for neutral LLPs using the endcap muon detectors was first published in 2021 [1]. To extend the technique used in the first analysis, a second paper [2] that uses both the barrel and the endcap muon detectors is recently submitted to Physical Review D. Additionally, the second paper also extended the interpretations of the twin Higgs scenario to lower LLP masses and additional decay modes, and added an interpretation to the dark shower model. This chapter focuses on presenting the latest result from the second paper [2] using both the barrel and endcap muon detectors.

This search is sensitive to the production of single or multiple LLPs decaying to final states including hadrons, τ leptons, electrons, or photons. The LLPs decaying to muons very rarely produce a particle shower and will generally not be detected by this search. While this search is sensitive to many models predicting LLPs, we interpret the results in two separate benchmark hidden sector scenarios. The first is a simplified model motivated by the twin Higgs scenario [98–100, 108–110] where the SM Higgs boson (H) decays to a pair of neutral long-lived scalars (S), each of which decays in turn to a pair of fermions or a pair of photons, as shown in Figure 5.1 (left). We search for long-lived scalars with masses between 0.4 and 55 GeV in a wide range of decay modes, including decays resulting primarily in hadronic showers ($b\bar{b}$, $d\bar{d}$, K^+K^-, K^0K^0, and $\pi^+\pi^-$), decays resulting primarily in electromagnetic showers ($\pi^0\pi^0$, $\gamma\gamma$, and e^+e^-), and decays to $\tau^+\tau^-$, which may result in hadronic or electromagnetic showers. The most stringent previous constraints for mean proper decay lengths $c\tau < 0.3$ m are based on a search for displaced jets in the CMS tracker [111]. For $c\tau > 0.3$ m, a search for displaced vertices in the ATLAS muon spectrometer [112, 113] set the most stringent previous limits.

We also interpret the search results in terms of a set of hidden sector "dark shower" models with perturbative parton showers [114]. We consider production of an SM Higgs boson that decays to a pair of dark-sector quarks, each of which hadronizes into a dark shower consisting of short- and long-lived dark-sector mesons (scalar or vector) that eventually decay back to SM particles. Depending on the symmetries and decay portal, the proper lifetime of the dark mesons and the final-state SM particles can vary, resulting in a wide range of dark-shower signatures. We interpret the search results in a framework for long-lived states with masses between 2–20 GeV and five different decay portals [114], namely the gluon portal producing hadron-rich showers, the photon portal with photon showers, the vector portal with semi-visible jets, the Higgs boson portal with heavy-flavor-rich showers, and the dark-photon portal with lepton-rich showers. This is the first search at the LHC with an interpretation in this framework. The diagram of this benchmark model is shown in Figure 5.1 (right).

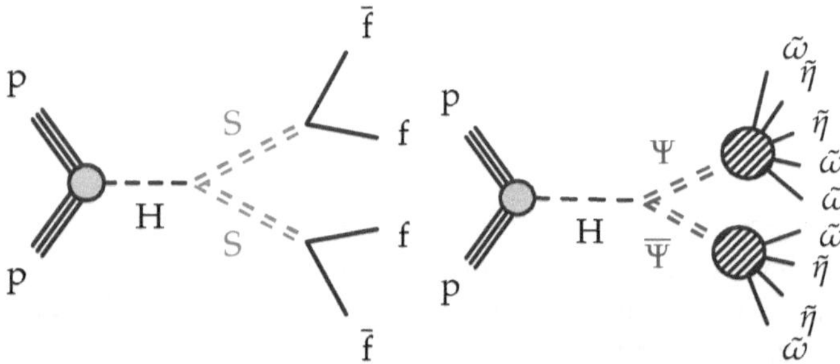

Figure 5.1: Diagrams of twin Higgs model (left) and dark-shower model (right). In the twin Higgs model, the SM Higgs boson (H) decays to a pair of neutral long-lived scalars (S), which then decay to two SM particles. Only fermions (f) are shown in the diagram, but the LLP may also decay to a pair of photons. In the dark-shower model, the H boson decays to a pair of dark-sector quarks (Ψ), which then hadronize to form dark showers consisting of dark scalar ($\tilde{\eta}$) and vector mesons ($\tilde{\omega}$) that decay back to SM particles.

There are two key advantages of the LLP search strategy presented in this chapter over searches that employ displaced vertices.

(i) The absorbers in front of the muon detectors act as shielding material to maintain a sufficiently low level of background for the detection of a single LLP decay. This level of background rejection could only be achieved in current hadronically decaying displaced-vertex searches by requiring the detection of two LLP decays.

(ii) The MDS signature is sensitive to the LLP energy only and insensitive to its mass, rendering this search equally sensitive to all LLP masses considered. In contrast, the vertex reconstruction efficiency in a displaced-vertex search tends to decrease with the LLP mass because of the increasingly smaller opening angles.

Because of these advantages, the signal acceptance and sensitivity are improved relative to the acceptance and sensitivity of analyses leading to the previous best results [111–113] for a wide range of LLP masses and proper lifetimes.

This chapter is organized as follows. Section 5.2 provides a summary of the data set and simulated samples used in the analysis. The reconstruction of the standard physics objects and muon detector showers are discussed in Section 5.3 and 5.4, respectively. The event selection is described in Section 5.5. The background estimation methods are detailed in Section 5.6. The signal modeling and systematic uncertainties are discussed in Sections 5.7. We report and interpret the results in Section 5.8. Finally, a summary is given in Section 5.9.

5.2 Dataset and Simulated Samples

5.2.1 Datasets and Triggers

The search uses proton-proton (pp) collision data collected at a center-of-mass energy of 13 TeV during 2016–2018 at the LHC, corresponding to an integrated luminosity of 138 fb^{-1}.

For signal scenarios where the LLP lifetime is relatively long, such that one or more LLPs decay beyond the calorimeters, a significant amount of missing transverse momentum (p_T^{miss}) will be produced as the LLP momentum remains undetected. In the benchmark twin Higgs signal model, for events where the Higgs boson is produced with significant recoil from initial state radiation (ISR), a large p_T^{miss} is produced. This feature is illustrated in the cartoon in Figure 5.2 and the correlation between the Higgs p_T and p_T^{miss} is shown in Figure 5.3. This analysis utilizes this feature by triggering on the p_T^{miss}-based triggers, with $p_T^{\mathrm{miss}} > 120$ GeV, selecting for a boosted Higgs phase space. Additionally, as will be described in Section 5.4, the reconstruction of the main object in this analysis, muon detector shower, requires the use of muon detector rechits that are stored only in the RECO data format. The RECO dataset is only available in the form of a p_T^{miss}-skim dataset due to the large event size of this data-tier, which has a $p_T^{\mathrm{miss}} > 200$ GeV requirement, well above the trigger threshold where the high-level trigger efficiency is effectively constant. The signal efficiency of the $p_T^{\mathrm{miss}} > 200$ GeV requirement is about 1% for benchmark signal models. A new dedicated trigger path for muon detector shower was developed and successfully commissioned for Run 3 that increases the signal efficiency by more than an order of magnitude, as detailed in Chapter 7.

5.2.2 Signal Simulation

The simulated H $\rightarrow$ SS signal samples are generated at next-to-leading order (NLO) with the POWHEG 2.0 [115–118] generator for the five main production processes: gluon fusion, vector boson fusion, associated production with a Z or W vector

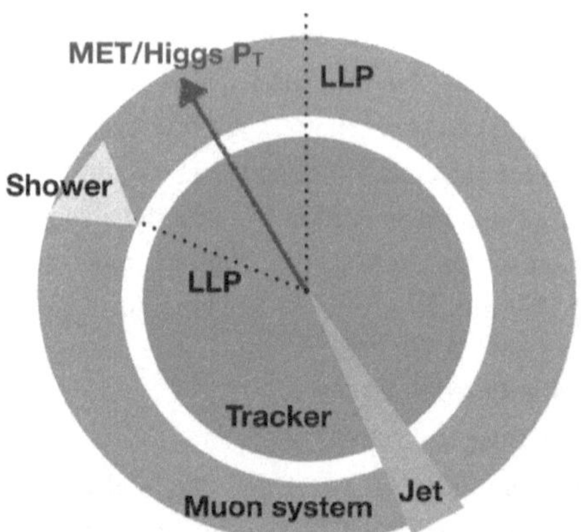

Figure 5.2: The $p_{\mathrm{T}}^{\mathrm{miss}}$ from signal comes from the recoil of the Higgs against ISR.

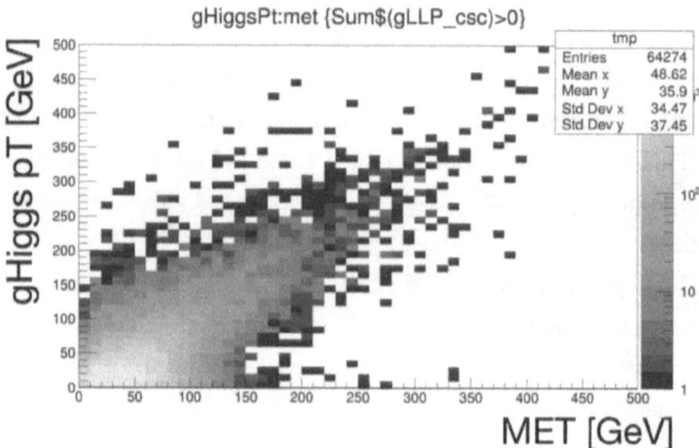

Figure 5.3: Strong correlation is observed between $p_{\mathrm{T}}^{\mathrm{miss}}$ and the generator-level Higgs p_{T} for the twin Higgs benchmark model.

boson, and associated production with a pair of top quarks. The Higgs boson mass is set to 125 GeV, while the S mass (m_{S}) is set to 0.4, 1.0, 1.5, 3.0, 7.0, 15.0, 40.0, or 55.0 GeV. The proper decay length is set to various values ranging between 1 mm and 100 m. We consider decays to $b\bar{b}$, $d\bar{d}$, K^+K^-, K^0K^0, and $\pi^+\pi^-$, hereafter referred to as the fully hadronic decay modes, decays to $\pi^0\pi^0$, $\gamma\gamma$, and e^+e^-, referred to as the fully electromagnetic decay modes, and decays to $\tau^+\tau^-$. Di-muon decays are not considered, since muons typically do not create high hit-multiplicity showers in the muon detectors. These specific decay modes are selected because they represent the dominant decay modes for Higgs-like scalar particles with mass in various ranges [119, 120]. For mass below $0.2\,\text{GeV}$, the e^+e^- and $\gamma\gamma$ decay modes are dominant; between 0.3 and $1\,\text{GeV}$, the $\pi^+\pi^-$ and $\pi^0\pi^0$ decay modes are dominant; between 1 and $2\,\text{GeV}$, the K^+K^- and K^0K^0 decay modes are dominant; and above $2\,\text{GeV}$ the $\gamma\gamma$ and qq decay modes are dominant. The generator-level p_T of the Higgs for gluon fusion production mode is reweighted to the best known theoretical prediction (NNLOPS) [121]. Due to the high p_T^{miss} requirement, the sub-dominant production modes provide a total of 65% increase to the gluon fusion-only signal yield at in the boosted Higgs phase space, as shown in Figure 5.4.

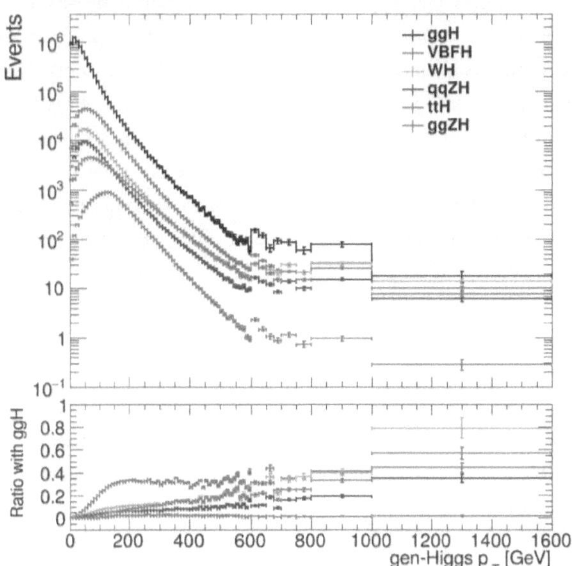

Figure 5.4: The generator-level Higgs p_T distribution for the different production modes of the Higgs boson and the ratio with gluon fusion is shown. The sub-dominant production modes provide non-neglible contribution to the signal yield at high Higgs pT. The gluon fusion distribution is reweighted to the best known theoretical prediction (NNLOPS) [121].

The dark shower signal models are generated similarly through Higgs boson production at NLO with POWHEG 2.0 and include only the dominant gluon fusion production mode. The Higgs boson mass is set to 125 GeV. The generator-level p_T of the Higgs for gluon fusion production mode is reweighted to the best known theoretical prediction (NNLOPS) [121]. The Higgs boson decay and the phenomenology of the dark showers are generated following the tools and theory priors presented in Ref. [114], using the PYTHIA 8 [122] hidden-valley module [123, 124]. In the generation, the dark sector is reduced to a single dark quark (Ψ), vector meson ($\widetilde{\omega}$), and scalar meson ($\widetilde{\eta}$), and there are three dark quantum chromodynamic (QCD) colors.

As mentioned in Section 5.1, we generate signal samples for five different decay portals. In the vector portal, $\widetilde{\omega}$ is long-lived and couples to SM particles, while $\widetilde{\eta}$ is invisible. For the dark-photon portal, $\widetilde{\eta}$ decays into a pair of long-lived dark photons with masses equal to $0.4m_{\widetilde{\eta}}$, which then each decay into SM particles. For all other portals, $\widetilde{\eta}$ is long-lived and couples to SM particles, while $\widetilde{\omega}$ is invisible. The LLP mass is varied between 2 and 20 GeV. The minimum LLP mass for each portal is motivated by the minimal ultraviolet completion and fine-tuning considerations discussed in Ref. [114]. The proper decay length is varied between 1 mm and 10 m. Characteristics of the models are the $\widetilde{\omega}$ to $\widetilde{\eta}$ meson mass ratio, ξ_ω, and the ratio of the dark sector QCD scale to the $\widetilde{\eta}$ mass, ξ_Λ. We consider three sets of numerical values: $(\xi_\omega, \xi_\Lambda) = (2.5, 2.5)$, $(2.5, 1.0)$, and $(1.0, 1.0)$. These three hierarchies represent regimes where $\widetilde{\omega}$ and $\widetilde{\eta}$ mesons are both produced and $\widetilde{\omega}$ can decay to a pair of $\widetilde{\eta}$ mesons, only $\widetilde{\eta}$ mesons are produced, and $\widetilde{\omega}$ and $\widetilde{\eta}$ mesons are both produced but $\widetilde{\omega}$ cannot decay to a pair of $\widetilde{\eta}$ mesons, respectively. The three sets of values are can be set for all portals, except for the vector portal, where only $\widetilde{\omega}$ couples to SM particles, so only $(\xi_\omega, \xi_\Lambda) = (1.0, 1.0)$ creates reconstructable signature. These distinct scenarios present a wide range of signatures with different LLP multiplicities, visible decay product multiplicities, and missing transverse momenta.

For both signal models, parton showering, hadronization, and the underlying event are modeled by PYTHIA 8.205 (8.230) [122] with parameters set by the CUETP8M1 [125] (CP5 [126]) tunes used for samples simulating the 2016 (2017 and 2018) data set conditions. The NNPDF3.0 [127] (3.1 [128]) parton distribution functions are used in the generation of all simulated samples. The GEANT4 [129] toolkit is used to model the response of the CMS detector. Simulated minimum-bias events are mixed with the hard interactions in simulated events to reproduce the effect of additional proton-proton interactions within the same or neighboring bunch crossings as the recorded event (pileup). Events are weighted such that the distribution of the number of interactions per bunch crossing agrees with that observed during each data-taking period.

5.3 Physics Objects

The particle-flow (PF) algorithm [75] is used to reconstruct and identify all the physics objects in an event, with details discussed in Section 4.2.7. Jet, muon, and p_T^{miss} are used in this search.

5.3.1 Jets

In this analysis, PF jets with $|\eta| < 2.4$ are used to veto background muon detector clusters caused by punch-through jets if the cluster centroid is within $\Delta R = \sqrt{(\Delta\eta)^2 + (\Delta\phi)^2} < 0.4$ of the PF jet. Jets with different p_T thresholds are used in different categories depending on the background rate in the categories, with more details discussed in Section 5.5.

Additionally, we use jets that pass the *tight lepton veto* identification criteria, defined in Table 5.1 and have $p_T > 30\,\text{GeV}$ and $|\eta| < 2.4$. The *tight lepton veto* working point is designed to remove jets originating from calorimetric noise and reject potential background from mis-reconstructed leptons. We require there to be at least one such jets in a signal event. This selection has a high signal efficiency, because the large p_T^{miss} requirement implies that a high p_T initial state radiation jet is recoiling against the Higgs boson in signal events.

	2016	2017 and 2018
neutral hadronic energy fraction	< 0.9	< 0.9
neutral electromagnetic energy fraction	< 0.9	< 0.9
number of constituents	> 1	> 1
muon fraction	< 0.8	< 0.8
charged hadron fraction	> 0	> 0
charged multiplicity	> 0	> 0
charged EM fraction	< 0.9	< 0.8

Table 5.1: *Tight lepton veto* jet identification selections for different years.

5.3.2 Muons

Standard muon reconstruction is used to construct muon candidates [67]. Muon objects with $|\eta| < 2.4$ are used to reject muon detector showers caused by muons producing a photon via bremsstrahlung if the cluster centroid is within $\Delta R < 0.4$ of the muon. Similar to the jet vetos, muons with different identification and p_T requirements are used in different categories depending on the muon rate in different categories and detector regions, with more details in Section 5.5.

5.3.3 Missing Transverse Momentum

The vector ($\vec{p}_\mathrm{T}^{\,\mathrm{miss}}$) is computed as the negative vector p_T sum of all the PF candidates in an event, and its magnitude is denoted as $p_\mathrm{T}^{\mathrm{miss}}$. The $\vec{p}_\mathrm{T}^{\,\mathrm{miss}}$ is modified to account for corrections to the energy scale of the reconstructed jets in the event [130]. In many cases, large $p_\mathrm{T}^{\mathrm{miss}}$ could come from unwanted noise, such as detector noise, cosmic rays, and beam-halo particles. Therefore, a series of filters are applied to remove the "fake" high-$p_\mathrm{T}^{\mathrm{miss}}$ events induced by the non-collisional background.

5.4 Muon Detector Showers

When charged particles traverse the CSCs, signal pulses are collected on the anode wires and the cathode strips. The signals from the wire groups are combined with signals from the cathode strips to form a point on a two-dimensional plane in each chamber layer called a CSC rechit. These hits are then used to form straight-line segments comprising at least three layers in the CSC chambers. When charged particles traverse the DT chambers, signal pulses are collected on the anode wires at the center of the DT cells. Since the DTs only provide measurement in either the ϕ or z dimension, the DT hit position is assumed to be at the center of each DT chamber in the orthogonal direction. The geometry and rechit reconstruction of DT and CSC chambers are detailed in Section 4.2.5. For LLPs that decay within or just in front of the muon system, the material in the iron return yoke structure will induce a hadronic or electromagnetic shower, creating a geometrically localized and isolated cluster of CSC and DT rechits in the muon detectors. The event display for an example simulated signal event is shown in Figure 5.5

During the initial R&D phase of the first endcap-only analysis, reconstructed CSC segments were studied as they were readily available in the more accessible AOD datasets, but the segment reconstruction algorithm results in saturation due to the excessively large number of hits in a small region of the detector. In Section 5.4.1, we show a detailed study on the use of CSC rechits as a proxy to identify signal-like showers. Section 5.4.2 details the reconstruction algorithm of the muon detector shower with CSC and DT rechits.

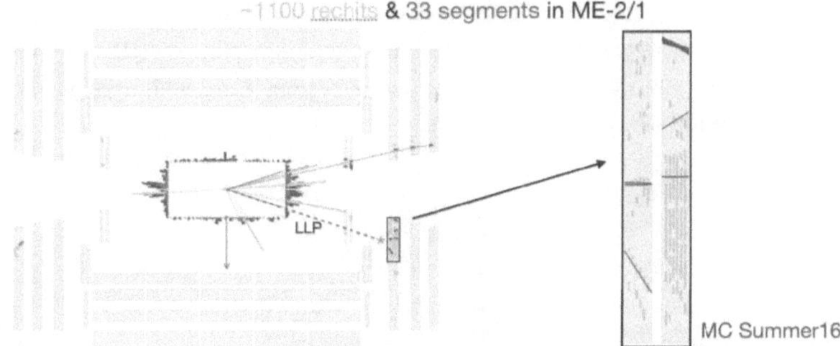

Figure 5.5: An event display of a simulated signal event from 2016. The LLP decays right in front of ME2/1, creating a large number of CSC rechits and segments in the chamber. CSC rechits are represented by the yellow dots and the CSC segments are represented by the purple lines in the muon end-cap region. There are 1100 rechits but only 33 segments reconstructed, demonstrating the use of rechits are more suitable for high-multiplicity shower reconstruction.

5.4.1 Using Rechits for Muon Detector Shower

This section details the study performed to validate the use of number of rechits as a proxy to detect the showers created by the LLP decays in the CSC system.

Initially, CSC segments were studied, as they are readily available in the more accessible AOD datasets. However, segment reconstruction algorithm is not designed to handle showers that have high multiplicity, as it is only designed to reconstruct segments from muons that leave only a few hits per layer (up to 6 hits per layer). Therefore, the large number of hits in the signal showers are not reflected in the number of segments. Furthermore, the CSC segment reconstruction for the 2017 and 2018 datasets had an additional beamspot contraint that significantly suppressed the efficiency for showering signals. Even for the segment reconstruction in 2016, as shown in the signal event display in Figure 5.5, 1100 CSC rechits were reconstructed for a signal shower, but only 33 segments are reconstructed. Therefore, we studied the use of rechits and its saturation for high-multiplicity showering signals and show that the rechits can be used to reflect the number of simulated hits from the signal showers and provide much better discrimination than the segments.

We studied the correlation between the number of simhits (N_{simhits}) and the number of rechits (N_{rechits}) for events where at least one LLP decayed in the CSC and observed a strong correlation between the two variables up to N_{simhits} of 600, as shown in Figure 5.6. We further observed that it is extremely rare for background events to have more than 600 simhits, as shown in Figure 5.7. Therefore, we use the number of rechits as a proxy to detect the showering signal with large number of simhits for this analysis.

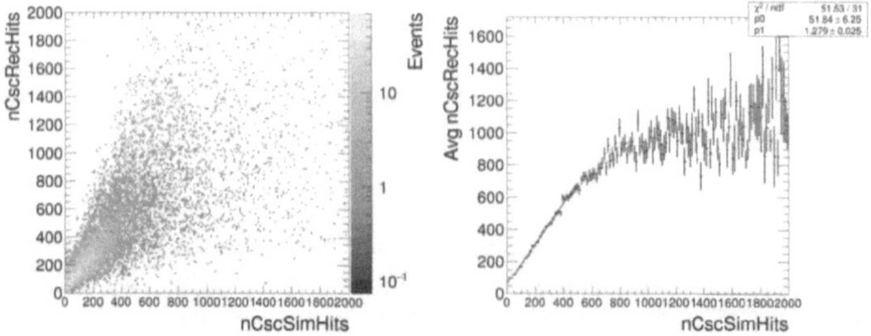

Figure 5.6: The correlation between the number of rechits and simhits (left) and the profileX of the 2D plot (right) are shown. The correlation between the number of simulated hits and reconstructed hits are linear up to $N_{\text{simhits}} = 600$.

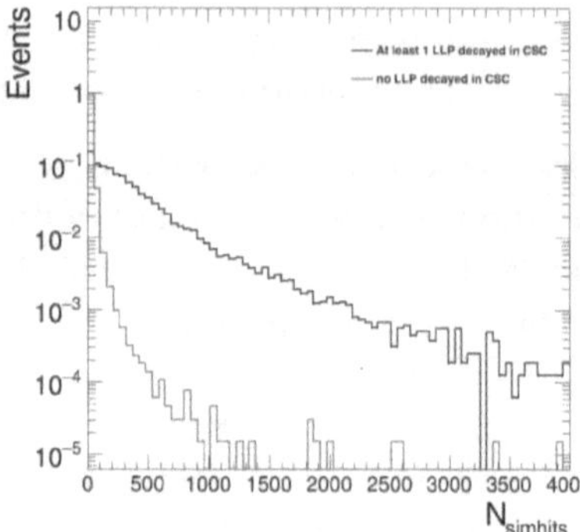

Figure 5.7: Number of simulated hits per event for events with at least one LLP decay in CSC (signal) and for events with LLP decay close to the interaction point (background-like events in the signal sample) are shown.

5.4.2 Muon Detector Shower Reconstruction

Because the CSC and DT hits contain different information, they are handled separately to reconstruct muon detector showers. The CSC and DT hits are clustered in η and the azimuthal angle ϕ (in radians) using the DBSCAN algorithm [131], which

groups hits by high-density regions. A minimum of 50 hits and a distance param-
eter $\Delta R_{\text{cluster}} = \sqrt{(\Delta\eta)^2 + (\Delta\phi)^2}$ of 0.2 is used. The minimum of 50 hits is chosen
because it is larger than the number of hits that a muon is expected to create (24
hits for CSC and 44 hits for DT) in the CSC or DT detectors. A spatial position is
associated with each cluster by taking the geometric center of the hits in the cluster.
From this, we can calculate the η and ϕ coordinates of each cluster. Nearby clusters
are merged if they satisfy $\Delta R < 0.6$. This is repeated until all clusters within an
event are isolated. This merging procedure ensures that clusters coming from the
same source are reconstructed as one object. In the overlap region of the muon
detectors, with $0.9 < |\eta| < 1.2$, if both CSC and DT clusters are reconstructed with
$\Delta R < 0.4$, the CSC clusters are given precedence because the CSC cluster response
is larger, and the corresponding DT clusters are removed.

The cluster reconstruction efficiency, including both DT and CSC clusters, as a
function of the simulated r and $|z|$ decay positions of the LLP for three different decay
modes are shown in Figure 5.8. The DT (CSC) cluster reconstruction efficiency is
shown separately as a function of the simulated r ($|z|$) for fully hadronic decays in
Figure 5.9. The efficiencies are shown for events satisfying $p_{\text{T}}^{\text{miss}} > 200\,\text{GeV}$, as
required in the search described in Section 5.2 and depends strongly on the LLP
decay position. The efficiency is highest when the LLP decays near the edges of the
shielding absorber material, where there is enough material to induce the shower,
but not so much that it stops the shower secondary particles. Decays that occur at
the beginning or just before a thick section of absorber material will have a reduced
efficiency because a fraction of the particle shower will be absorbed by the material
before it can be detected. The efficiency decreases to zero when the decay occurs
near the end of the last stations in MB4 or ME4 because there is an insufficient
amount of absorber material to induce any particle shower.

The cluster reconstruction efficiency also depends on whether the LLP decays to
hadrons or to electrons or photons. In general, hadronic showers have higher
efficiency compared to electromagnetic showers because they are more likely to
penetrate through the steel in between the muon stations. Showers induced by
electromagnetic decays generally occupy just one station and are stopped by the
steel between the stations. When the LLP decays close to or in the CSCs, defined
as the union of the two regions: (i) $500 < |z| < 661\,\text{cm}$, $r < 270\,\text{cm}$, and (ii)
$|\eta| < 2.4$ and (ii) $660 < |z| < 1100\,\text{cm}$, $r < 695.5\,\text{cm}$, and $|\eta| < 2.4$, the inclusive
CSC cluster reconstruction efficiency is approximately 80%, 55%, and 35% for fully

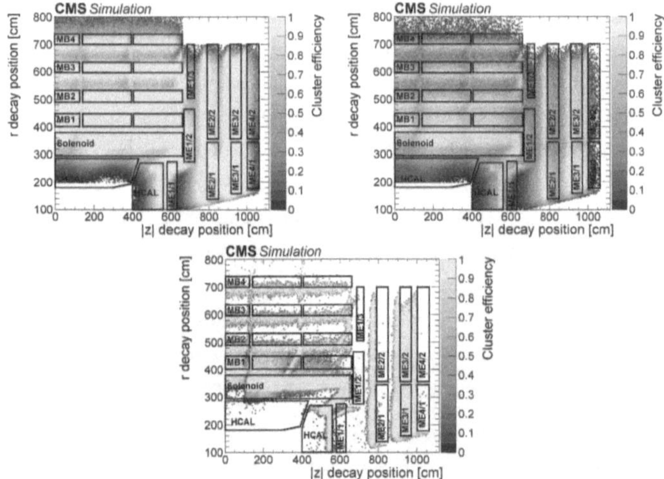

Figure 5.8: The cluster reconstruction efficiency, including both DT and CSC clusters, as a function of the simulated r and $|z|$ decay positions of the particle S decaying to $d\bar{d}$ (top left) , $\tau^+\tau^-$ (top right), and e^+e^- (bottom) in events with $p_T^{miss} > 200\,\mathrm{GeV}$. For $d\bar{d}$ and $\tau^+\tau^-$ the LLP mass is $40\,\mathrm{GeV}$ with a range of $c\tau$ values between 1 and $10\,\mathrm{m}$. For e^+e^- the LLP mass is $0.4\,\mathrm{GeV}$ with a range of $c\tau$ values between 0.01 and $1\,\mathrm{m}$. The cluster reconstruction efficiency appears to be nonzero beyond MB4 because the MB4 chambers are staggered so that the outer radius of the CMS detector ranges from 738 to $800\,\mathrm{cm}$. The barrel and endcap muon stations are drawn as black boxes and labeled by their station names. The region between labeled sections are mostly steel return yoke.

hadronic, $\tau^+\tau^-$, and fully leptonic decays, respectively. When the LLP decays close to or in the DTs, defined as the region $380 < r < 738\,\mathrm{cm}$ and $|z| < 661\,\mathrm{cm}$, the inclusive DT cluster reconstruction efficiency is approximately 80%, 60%, and 45% for fully hadronic, $\tau^+\tau^-$, and fully leptonic decays, respectively.

The accuracy of the simulation modeling of the cluster reconstruction efficiency has been studied using a $Z \rightarrow \mu^+\mu^-$ data sample, where clusters are produced when one of the muons undergoes bremsstrahlung and the associated photon produces an electromagnetic shower. The discrepancy between data and simulation is taken as a systematic uncertainty in the cluster reconstruction efficiency, as detailed in Section 5.7.

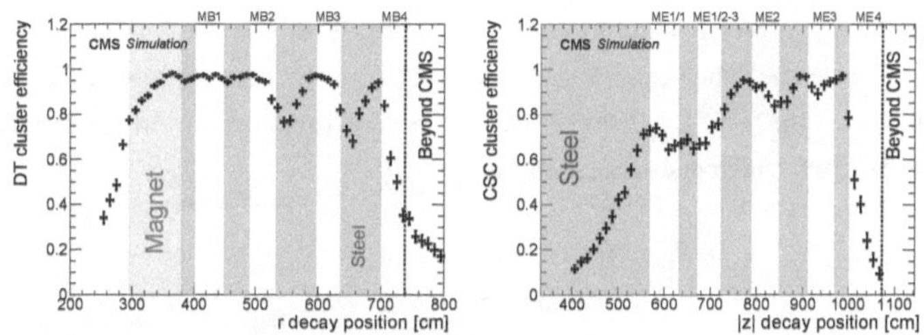

Figure 5.9: The DT (left) and CSC (right) cluster reconstruction efficiency as a function of the simulated r or $|z|$ decay positions of S decaying to $\mathrm{d\bar{d}}$ in events with $p_\mathrm{T}^\mathrm{miss} > 200\,\mathrm{GeV}$, for a mass of $40\,\mathrm{GeV}$ and a range of $c\tau$ values between 1 and 10 m. The DT cluster reconstruction efficiency is shown for events where the LLP decay occurs at $|z| < 700\,\mathrm{cm}$. The DT cluster reconstruction efficiency appears to be nonzero beyond MB4 because the MB4 chambers are staggered so that the outer radius of the CMS detector ranges from 738 to 800 cm. The CSC cluster reconstruction efficiency is shown for events where the LLP decay occurs at $|r| < 700\,\mathrm{cm}$ and $|\eta| < 2.6$. Regions occupied by steel shielding are shaded in gray.

5.5 Analysis Strategy

An LLP, like the neutral scalar or the dark meson, that decays after it has traversed the calorimeter systems may produce large $p_\mathrm{T}^\mathrm{miss}$ because its momentum is not properly measured or properly associated with a particle by the PF algorithm. We exploit this feature by analyzing the data collected by triggering on events with online $p_\mathrm{T}^\mathrm{miss} > 120\,\mathrm{GeV}$ [132], and subsequently requiring offline $p_\mathrm{T}^\mathrm{miss} > 200\,\mathrm{GeV}$, to ensure that the selected events are well above the trigger threshold where the high-level trigger efficiency is effectively constant. We require at least one jet with $p_\mathrm{T} > 30\,\mathrm{GeV}$ and $|\eta| < 2.4$, because signal events passing the $p_\mathrm{T}^\mathrm{miss}$ requirement are always produced together with a jet from initial-state radiation. The event-level selections are kept minimal to be as model independent as possible, and the efficiency of these selections is about 95%.

After the event-level selections, the events are separated into three mutually exclusive categories based on the number and location of the clusters: events (1) with two clusters in the muon detectors, (2) with exactly one CSC cluster, and (3) with exactly one DT cluster. Events with two clusters are further categorized into categories with two CSC clusters, two DT clusters, and one CSC and one DT cluster. The category with exactly one CSC cluster is based on a previous search using the endcap muon detectors [1] with a few minor changes, such as explicitly excluding overlapping events with two clusters and loosening the event-level selections to be consistent with those in the other categories. The geometric acceptance multiplied by the efficiency of the $p_T^{miss} > 200\,\text{GeV}$ selection as a function of LLP proper decay length for each category is shown in Figure 5.10.

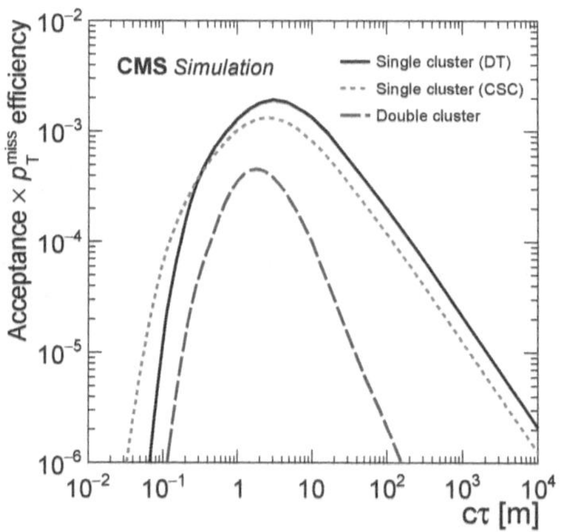

Figure 5.10: The geometric acceptance multiplied by the efficiency of the $p_T^{miss} > 200\,\text{GeV}$ selection as a function of the proper decay length $c\tau$ for a scalar particle S with a mass of 40 GeV.

The main SM backgrounds are similar among the three categories and include punch-through jets, muons that undergo bremsstrahlung, and isolated hadrons from pileup, recoils, or underlying events. To suppress background from punch-through jets or muon bremsstrahlung, we reject CSC and DT clusters that have a jet or muon within $\Delta R < 0.4$ in all categories. However, depending on the category, the p_T thresholds and identification requirements of jets and muons are different. Furthermore, additional tighter vetoes are applied to the single-cluster categories to reject background, as discussed in the following subsections.

Additionally, the angular difference ($\Delta\phi(\vec{p}_T^{\,miss}, \text{cluster})$) between the $\vec{p}_T^{\,miss}$ and the cluster location is used as a discriminating variable in all three categories. For signal, $\Delta\phi(\vec{p}_T^{\,miss}, \text{cluster})$ peaks near zero because the large p_T^{miss} requirement tends to select events where the Higgs boson has a large momentum and is nearly collinear with the S. For the backgrounds, $\Delta\phi(\vec{p}_T^{\,miss}, \text{cluster})$ is uniformly distributed because the cluster and $\vec{p}_T^{\,miss}$ are independent. The exact threshold on the variable is different for the barrel and endcap, and looser for the double-cluster category, as detailed in the following subsections.

Finally, the number of hits in the clusters (N_{hits}) is used to discriminate signal and background. Signal clusters tend to have large N_{hits}, while background clusters are expected to have small N_{hits}. The exact threshold on N_{hits} is different for the barrel and endcap, and looser for the double-cluster category, as detailed in the following subsections that describe the detailed event selections for each category.

5.5.1 Double Clusters

The double-cluster category includes events containing two clusters satisfying the selection criteria described below. Events are separated into three categories depending on whether there are two DT clusters, two CSC clusters, or one CSC and one DT cluster present. Requiring two muon system clusters significantly reduces the expected background, so the selection requirements in the double-cluster category are much looser compared to the single cluster categories.

To reject clusters from punch-through jets and muon bremsstrahlung, the CSC clusters are rejected if any jet with $p_{\text{T}} > 30\,\text{GeV}$ or a global muon [67], with $p_{\text{T}} > 30\,\text{GeV}$ is found within $\Delta R < 0.4$. Similarly, DT clusters are rejected if any jet with $p_{\text{T}} > 50\,\text{GeV}$ or a muon passing loose identification criteria [67, 133] with $p_{\text{T}} > 10\,\text{GeV}$ is found within $\Delta R < 0.4$.

We veto CSC clusters that are entirely contained in the innermost rings of the ME1 station (ME1/1) and veto DT clusters that have more than 90% of the hits contained in the innermost station (MB1), both of which have the least absorber material between them and the interaction point, and larger background contamination.

We further reject CSC clusters produced by adjacent bunch crossings, known as out-of-time (OOT) pileup, by requiring that the cluster time (t_{cluster}) is consistent with an in-time interaction ($-5.0 < t_{\text{cluster}} < 12.5\,\text{ns}$). The cluster time is defined as the average time of the hits in the cluster relative to the LHC clock and corrected for the particle time-of-flight from the interaction point to the respective muon detector chambers assuming speed of light propagation. An asymmetric time window is used to capture signal clusters with longer delays from slower-moving LLPs. To reject clusters composed of hits from multiple bunch crossings, the root-mean-square spead of the hit times of each cluster is required to be less than 20 ns.

Tracks from muons in the barrel are likely to deposit a similar number of hits in all four DT stations, while showers from LLP decays are likely to have hits concentrated in one or two stations. Therefore, to reject DT clusters from muon bremsstrahlung we veto clusters that contain hits in all four stations and that have a ratio of the minimum and maximum number of hits per station less than 0.4.

Cosmic ray muons produce hits in both the upper and lower hemispheres of the muon barrel system. To suppress this background, we reject DT clusters if there are at least six segments, which are straight-line tracks built within each DT chamber, and at least one segment in every station found in the opposite hemisphere ($|\Delta\phi| > 2$) from the cluster. In addition, cosmic ray muon showers produce hits in multiple regions of the CMS detector. Thus, we reject any event in which more than a quarter of the DT or CSC rings, consisting of chambers with the same r and z coordinates, contain 50 or more hits. Finally, we require $\Delta\phi(\vec{p}_\mathrm{T}^\mathrm{miss}, \mathrm{cluster}) < 1.0 \ (1.2)$ for CSC (DT) clusters.

For signal events with two clusters, the ΔR between two LLPs, thus between the clusters, is typically small, due to the high $p_\mathrm{T}^\mathrm{miss}$ selection. For background, the ΔR between clusters is usually large because the two clusters generally come from separate processes, especially for the CSC-CSC and DT-CSC subcategories. Therefore, we require the ΔR between the two clusters to be less than 2 (2.5) for the CSC-CSC (DT-CSC) subcategory. There is already an implicit $\Delta\phi$ selection between the two clusters by requiring both clusters to pass the $\Delta\phi(\vec{p}_\mathrm{T}^\mathrm{miss}, \mathrm{cluster})$ selection. The ΔR selection additionally requires the two clusters to be close in η.

Finally, N_hits is used to discriminate signal and background. We require $N_\mathrm{hits} \geq 100$ (80) for CSC (DT) clusters.

5.5.2 Single CSC Cluster

The single-CSC-cluster category includes events in which only one LLP decay produces a displaced cluster in the endcap muon system. In this category, the expected background yield is significantly higher than in the double-cluster category, so we apply much tighter cluster veto requirements to achieve the same near-zero background level.

The cluster veto requirements are the same as in the first endcap-only search [1], except the event-level selections are loosened to align with the other two categories, and events with two clusters are only included in the double-cluster category, which has higher sensitivity. Clusters that have a jet with $p_\mathrm{T} > 10\,\mathrm{GeV}$ or a muon with $p_\mathrm{T} > 20\,\mathrm{GeV}$ within $\Delta R < 0.4$ are rejected. We additionally veto clusters that have any hits in the two innermost rings of the ME1 station (ME1/1 and ME1/2), which have the least absorber material between them and the interaction point, or are matched to any hits (with $\Delta R(\text{cluster, hit}) < 0.4$) in the RPCs located immediately next to ME1/2 (RE1/2). In the region where the barrel and endcap muon detectors overlap ($0.9 < |\eta| < 1.2$), any cluster that is matched to any DT segments in MB1 or any hit in RB1 within $\Delta R(\text{cluster, segment or hit}) < 0.4$, is vetoed. We reject clusters with $|\eta| > 2.0$ to suppress the muon bremsstrahlung background that evaded the muon veto because of the decreasing muon reconstruction and identification efficiencies at larger $|\eta|$. Finally, as in the double-cluster category, events in which more than a quarter of the DT or CSC rings contain 50 or more hits are rejected.

After the veto requirements are applied, the dominant background source consists of decays of low p_T SM LLPs, which are predominantly produced by pileup interactions and are independent of the primary interaction that yielded the large p_T^{miss}. These pileup interactions may be in-time or OOT with the primary interaction, as shown in Figure 5.11. Clusters produced by OOT pileup are rejected by requiring $-5.0 < t_{\text{cluster}} < 12.5$ ns, as in the double-cluster category. The time window requirement suppresses the background by a factor of 5. Similarly, the root-mean-square spread of the hit times of each cluster is required to be less than 20 ns. The low p_T particle background components are further studied extensively with minimum bias simulation samples and particle gun samples, as detailed in Appendix 5.A.

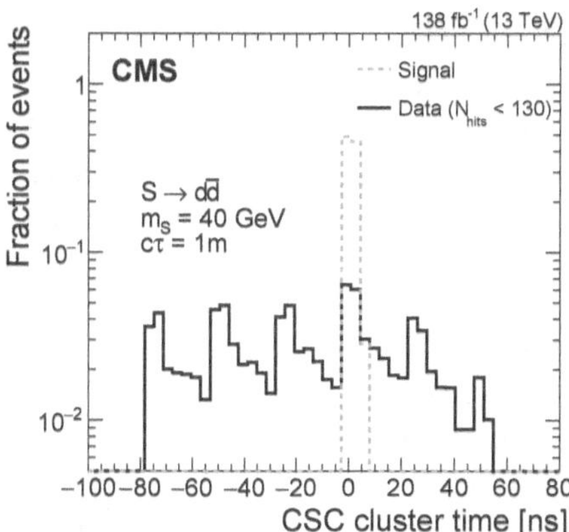

Figure 5.11: The shapes of the cluster time for signal, where S decaying to $d\bar{d}$ for a proper decay length $c\tau$ of 1 m and mass of 40 GeV, and for a background-enriched sample in data selected by inverting the N_{hits} requirement.

To distinguish signal and background clusters it was observed that clusters from all background processes occur more often at larger values of $|\eta|$, as the effectiveness of the jet and muon vetoes decreases because of decreasing reconstruction efficiencies. Signal clusters often occupy more than one CSC station ($N_{\text{stations}} > 1$) and occur more frequently in stations farther away from the primary interaction point. A cluster identification algorithm was devised that makes more restrictive $|\eta|$ requirements for clusters that occupy only one CSC station ($N_{\text{stations}} = 1$) and are closer to the primary interaction point. The $|\eta|$ requirements are:

- $|\eta| < 1.9$ if $N_{\text{stations}} > 1$,

- $|\eta| < 1.8$ if $N_{\text{stations}} = 1$ and the cluster is in station 4,

- $|\eta| < 1.6$ if $N_{\text{stations}} = 1$ and the cluster is in station 3 or station 2, and

- $|\eta| < 1.1$ if $N_{\text{stations}} = 1$ and the cluster is in station 1 because of an implicit selection from the ME1/1 and ME1/2 vetoes.

The cluster identification algorithm has $\sim 80\%$ efficiency for simulated clusters originating from S decays, and suppresses the background by a factor of 3. The events that pass the cluster identification criteria are used to define the search region (or signal region, SR), and those that fail are used as an in-time validation region (VR) for background estimation.

Both N_{hits} and $\Delta\phi(\vec{p}_{\text{T}}^{\text{miss}},\text{cluster})$ are used to discriminate signal and background. The signal and background shapes of the two discriminants are shown in Figure 5.12. For the backgrounds, $\Delta\phi(\vec{p}_{\text{T}}^{\text{miss}},\text{cluster})$ is independent of N_{hits}, enabling the use of the ABCD method to predict the background yield in the signal-enriched bin, as detailed in Section 5.6.

5.5.3 Single DT Cluster

The single-DT-cluster category targets events in which only one LLP decay produces a displaced cluster in the barrel muon system. Events passing the selection criteria for the double-cluster or single-CSC-cluster categories are not considered in this category, to give precedence to the category with higher sensitivity (double cluster) and to minimize differences with the previously published search (single CSC cluster).

First, to remove high-$p_{\text{T}}^{\text{miss}}$ events due to mismeasured jets, we require the minimum of $|\Delta\phi(\text{jet}, \vec{p}_{\text{T}}^{\text{miss}})|$ over all the jets with $p_{\text{T}} > 30\,\text{GeV}$ to be greater than 0.6. This

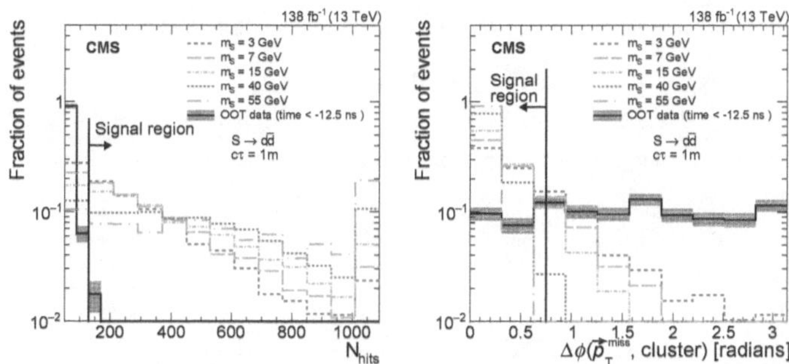

Figure 5.12: The shapes of N_{hits} (left) and $\Delta\phi(\vec{p}_T^{\text{miss}}, \text{cluster})$ (right) for single CSC clusters are shown for S decaying to $d\bar{d}$ for a proper decay length of 1 m and various masses compared to the OOT background ($t_{\text{cluster}} < -12.5$ ns). The OOT background is representative of the overall background shape, because the background passing all the selections described above is dominated by pileup and underlying events. The shaded bands show the statistical uncertainty in the background.

requirement reduces the background from SM events composed uniquely of jets produced through the strong interaction, referred to as QCD multijet events, and is only applied to the single-DT-cluster category because this category is dominated by the punch-through jet background.

We veto clusters that have a jet with $p_T > 10\,\text{GeV}$ or a muon with $p_T > 10\,\text{GeV}$ passing loose identification criteria [67, 133] within $\Delta R < 0.4$. In addition, we reject clusters that are within $\Delta R < 1.2$ from the leading-p_T jet. Furthermore, DT clusters that are within $\Delta R < 0.4$ of two or more hits in the innermost station MB1 are rejected. Additionally, clusters with maximum hit counts in MB3 or MB4 are rejected if they are within $\Delta R < 0.4$ of two or more MB2 hits. Each cluster is associated with one of the five wheels based on the average z position of its hits. To reject clusters from noise in the DTs, we require clusters to be matched to at least one RPC hit from the same wheel and within $\Delta\phi < 0.5$.

To suppress background from cosmic ray muons, we veto clusters that have more than 8 hits in MB1 within $\Delta\phi < \pi/4$ in either adjacent wheel. In addition, we veto clusters with maximum hit counts in MB3 and MB4 that have more than 8 hits in MB2 within $\Delta\phi < \pi/4$ in either adjacent wheel. Furthermore, we look for DT segments that are far from the clusters with $\Delta R > 0.4$ in the upper or lower hemisphere of the DT wheels. We veto the cluster if more than 14 segments are found in either hemisphere or more than 10 segments are found in both hemispheres.

As in the single-CSC-cluster category, N_{hits} and $\Delta\phi(\vec{p}_{\text{T}}^{\text{miss}}, \text{cluster})$ are used to discriminate signal and background. We require $\Delta\phi(\vec{p}_{\text{T}}^{\text{miss}}, \text{cluster}) < 1$ and $N_{\text{hits}} > 100$. The signal and background distributions of the two discriminants are shown in Figure 5.13.

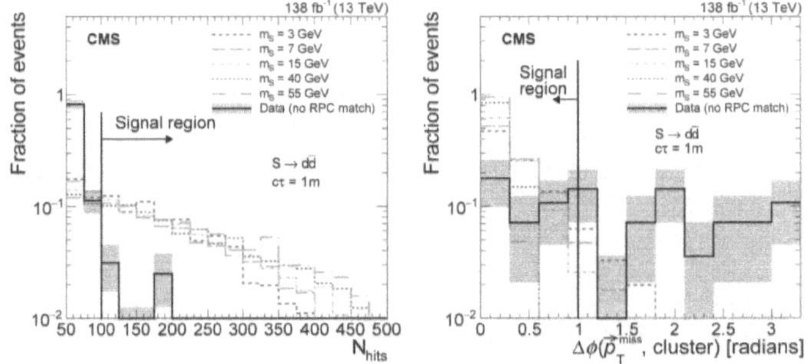

Figure 5.13: The shapes of N_{hits} (left) and $\Delta\phi(\vec{p}_{\text{T}}^{\text{miss}}, \text{cluster})$ (right) for DT clusters are shown for S decaying to $d\bar{d}$ for a proper decay length of 1 m and various masses compared to the shape of background in a selection in which the cluster is not matched to any RPC hit. The shaded bands show the statistical uncertainty in the background.

Finally, the DT clusters are categorized into 3 exclusive categories according to the station that contains the most hits: MB2, MB3, or MB4. These categories have different background compositions, where the punch-through jet background is more prominent in the stations that are closer to the interaction point.

5.6 Background Estimation

The "ABCD" method [134] based on control samples in data is used for background estimation for all three categories. The ABCD method requires two variables that discriminate signal and background and are independent of one another for the background. Two separate requirements, one on each variable, partition the two-dimensional space into four bins, A, B, C, and D, where bin A contains events that pass both signal-like requirements, events in bins B and D only pass one of the requirements, and events in bin C pass neither requirement. Because of the independence of the two variables, the expected background event rate in the signal-enriched bin A is related to the other three bins by $\lambda_{\text{A}} = (\lambda_{\text{B}}\lambda_{\text{D}})/\lambda_{\text{C}}$, where λ_{X} is the expected background event rate (i.e., Poisson mean) in each bin X. In the double-cluster categories, the two variables that are used are the N_{hits} of each of the clusters, while in the single DT and CSC cluster categories, the two variables are N_{hits} and $\Delta\phi(\vec{p}_{\text{T}}^{\,\text{miss}}, \text{cluster})$. To account for a potential signal contribution to bins A, B, and C, a binned maximum likelihood fit is performed simultaneously in the four bins, with a common signal strength parameter scaling the signal yields in each bin. The background component of the fit is constrained to obey the ABCD relationship.

In addition to the background predicted by the ABCD method, the single-DT-cluster category estimates the punch-through jet background separately, while the punch-through jet background in the other categories is negligible. Other non-collision backgrounds, including cosmic ray muons, have been suppressed by dedicated filters described in Section 5.5 and have been demonstrated to be negligible in the SR. In particular the background components in the single CSC category is dominated by low p_{T} particles from underlying events, which has been further studied with minimum bias simulation samples and particle gun samples, as detailed in Appendix 5.A.

The following subsections describe the main background component for each category, the background estimation method, and its validation.

5.6.1 Double Clusters

For the DT-CSC category, the two independent discriminating variables are the N_{hits} of the DT and CSC cluster, respectively. The four bins comprise events with clusters having the following properties, and are shown in Figure 5.14 (left):

- Bin A: CSC cluster with $N_{\text{hits}} > 100$ and DT cluster with $N_{\text{hits}} > 80$;

- Bin B: CSC cluster with $N_{\text{hits}} > 100$ and DT cluster with $N_{\text{hits}} \leq 80$;

- Bin C: CSC cluster with $N_{\text{hits}} \leq 100$ and DT cluster with $N_{\text{hits}} \leq 80$; and

- Bin D: CSC cluster with $N_{\text{hits}} \leq 100$ and DT cluster with $N_{\text{hits}} > 80$.

The estimation of the number of events in each bin is expressed by:

$$
\begin{aligned}
N_A &= \lambda_{\text{B}} \times \lambda_{\text{D}} / \lambda_{\text{C}} + \mu \times Sig_A \\
N_B &= \lambda_{\text{B}} + \mu \times Sig_B \\
N_C &= \lambda_{\text{C}} + \mu \times Sig_C \\
N_D &= \lambda_{\text{D}} + \mu \times Sig_D
\end{aligned}
\tag{5.1}
$$

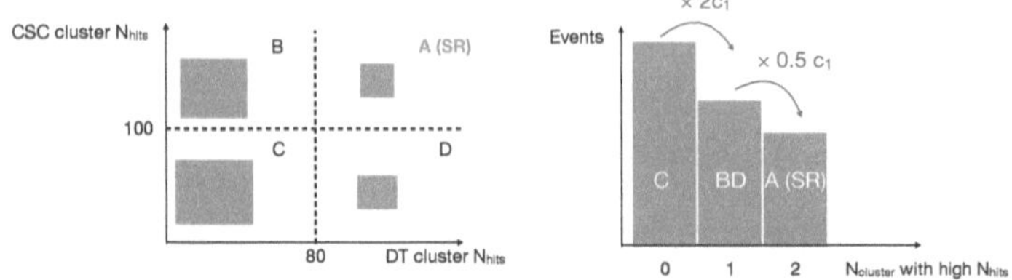

Figure 5.14: Diagrams illustrating the ABCD plane for the DT-CSC category (left), and for the DT-DT and CSC-CSC categories (right). The variable c_1 is the pass-fail ratio of the N_{hits} selection for the background cluster. Bin A is the SR for all categories.

where:

- Sig_A, Sig_B, Sig_C, and Sig_D are the number of signal events expected in bin A, B, C, and D, predicted by the signal MC simulation sample,

- μ is the signal strength (the parameter of interest of the model), and

- λ_B, λ_C, and λ_D are the number of background events in bin B, C, and D. They are interpreted as nuisance parameters that are unconstrained in the fit.

The four unknown variables (λ_B, λ_C, λ_D, and μ) are extracted from a maximum likelihood fit with the following likelihood expression:

$$L = \prod_i^{ABCD} Pois(obs_i|\lambda_i) \times \prod_i^{nuisance} Constraints(\sigma_j|\hat{\sigma}_j) \tag{5.2}$$

where obs_i is the number of observed events in each bin and σ_j are the nuisance parameters that capture the impact of systematic uncertainties. The MC statistics are implemented with Gamma distribution [135]. All other nuisance parameters are implemented with log-normal distributions, such that the logarithm of the distribution is a Gaussian constraint.

For the CSC-CSC and DT-DT categories, the two variables are symmetric, so we combine bins B and D and define the combined expected background rate as λ_{BD}. Bins A, BD, and C contain events with 2, 1, and 0 clusters passing the N_{hits} selection, respectively, as shown in Figure 5.14 (right). The expected background yield in the signal-enriched bin A is related to the other two bins as $\lambda_A = (\lambda_{\text{BD}}/2)^2/\lambda_C$.

The estimation of the number of events in each bin is expressed by:

$$N_A = (\lambda_{\mathrm{BD}}/2)^2/\lambda_{\mathrm{C}} + \mu \times Sig_A$$
$$N_{\mathrm{BD}} = \lambda_{\mathrm{BD}} + \mu \times Sig_{\mathrm{BD}}$$
$$N_C = \lambda_{\mathrm{C}} + \mu \times Sig_C \qquad (5.3)$$

where:

- Sig_A, $Sig_B D$, and Sig_C are the number of signal events expected in bin A, BD, and C, predicted by the signal MC simulation sample,

- μ is the signal strength (the parameter of interest of the model), and

- λ_{BD} and λ_{C} are the number of background events in bin BD and C. They are interpreted as nuisance parameters that are unconstrained in the fit.

The three unknown variables (λ_{BD}, λ_{C}, and μ) are extracted from a maximum likelihood fit with the following likelihood expression:

$$L = \prod_i^{A,BD,C} Pois(obs_i|\lambda_i) \times \prod_i^{nuisance} Constraints(\sigma_j|\hat{\sigma}_j) \qquad (5.4)$$

where obs_i is the number of observed events in each bin and σ_j are the nuisance parameters that capture the impact of systematic uncertainties. Similar to the DT-CSC category, the MC statistics are implemented with Gamma distribution and all other nuisance parameters are implemented with log-normal distributions.

To validate the background estimation method for the double cluster categories, we define two validation regions (VRs): the inverted-N_{hits} region and the inverted-$\Delta\phi(\vec{p}_{\mathrm{T}}^{\,\mathrm{miss}},\mathrm{cluster})$ region. The inverted-N_{hits} VR is defined by inverting the N_{hits} requirements for both clusters, while maintaining all the other cluster-level selections. The N_{hits} threshold used in the inverted-N_{hits} VR is 70 (80) for DT (CSC) clusters. Similarly, the inverted-$\Delta\phi(\vec{p}_{\mathrm{T}}^{\,\mathrm{miss}},\mathrm{cluster})$ VR is defined by inverting the $\Delta\phi(\vec{p}_{\mathrm{T}}^{\,\mathrm{miss}},\mathrm{cluster})$ requirement of both clusters while maintaining all the other cluster-level selections. To probe signal-like events in the inverted-$\Delta\phi(\vec{p}_{\mathrm{T}}^{\,\mathrm{miss}},\mathrm{cluster})$ VR, we additionally require $\Delta\phi(\mathrm{cluster1},\mathrm{cluster2}) < 2$ for the DT-DT category. For the CSC-CSC and DT-CSC categories, the two clusters are close to each other because of the $\Delta R(\mathrm{cluster1},\mathrm{cluster2})$ requirement. The $\Delta\phi(\vec{p}_{\mathrm{T}}^{\,\mathrm{miss}},\mathrm{cluster})$ VR allows us to test for any nonnegligible backgrounds at high N_{hits} that cannot be accessed in the inverted-N_{hits} VR. The background estimate agrees with the number of observed background events in both VRs and all three categories, as shown in Table 5.2.

Table 5.2: Validation of the ABCD method for the double-cluster category in both VRs. The uncertainty in the prediction is the statistical uncertainty propagated from bins B, C, and D or bins BD and C. The symbol λ_X is the expected background event rate in bin X, while N_X is the observed number of events in bin X.

Category	VR	λ_A	N_A	N_B	N_C	N_D	N_{BD}
DT-DT	Inverted $\Delta\phi(\vec{p}_T^{\,miss}, \text{cluster})$	0.02 ± 0.05	0	—	11	—	1
	Inverted N_{hits}	0.12 ± 0.27	0	—	2	—	1
CSC-CSC	Inverted $\Delta\phi(\vec{p}_T^{\,miss}, \text{cluster})$	0.12 ± 0.18	0	—	8	—	2
	Inverted N_{hits}	0.25 ± 0.38	0	—	4	—	2
DT-CSC	Inverted $\Delta\phi(\vec{p}_T^{\,miss}, \text{cluster})$	0 ± 0.3	0	0	19	3	—
	Inverted N_{hits}	0.18 ± 0.23	0	2	11	1	—

5.6.2 Single CSC Cluster

For the single-CSC-cluster category, the two discriminating variables are $\Delta\phi(\vec{p}_T^{\,miss}, \text{cluster})$ and N_{hits}. For the backgrounds, $\Delta\phi(\vec{p}_T^{\,miss}, \text{cluster})$ is independent of N_{hits} enabling the use of the ABCD method. As shown in Figure 5.15, the four bins comprise events with clusters having the following properties:

- Bin A: $\Delta\phi(\vec{p}_T^{\,miss}, \text{cluster}) < 0.75$ and $N_{hits} > 130$;

- Bin B: $\Delta\phi(\vec{p}_T^{\,miss}, \text{cluster}) \geq 0.75$ and $N_{hits} > 130$;

- Bin C: $\Delta\phi(\vec{p}_T^{\,miss}, \text{cluster}) \geq 0.75$ and $N_{hits} \leq 130$; and

- Bin D: $\Delta\phi(\vec{p}_T^{\,miss}, \text{cluster}) < 0.75$ and $N_{hits} \leq 130$.

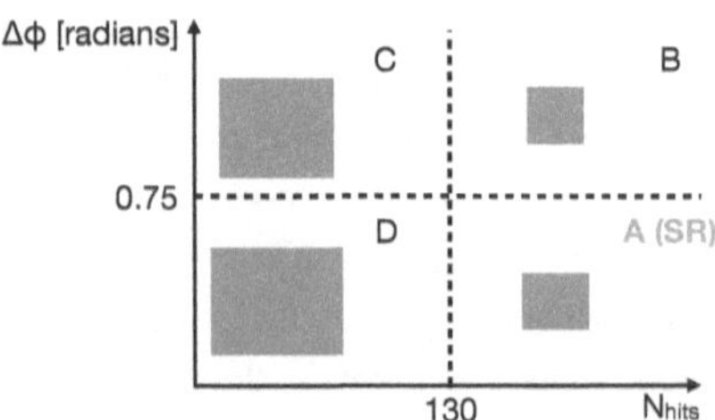

Figure 5.15: Diagram illustrating the ABCD plane for the single-CSC-cluster category, where bin A is the SR.

The estimation of the number of events in each bin is the same as that of the DT-CSC category expressed in Equation 5.1. The background estimation procedure is validated using events in two separate VRs: one in the OOT region, where $t_{\text{cluster}} < -12.5$ ns, and one in the in-time region, where the clusters that fail the cluster identification criteria are selected, as shown in Table 5.3.

Table 5.3: Validation of the ABCD method for the single-CSC-cluster category in both VRs. The uncertainty in the prediction is the statistical uncertainty propagated from bins B, C, and D. The symbol λ_X is the expected background event rate in bin X, while N_X is the observed number of events in bin X.

VR	λ_A	N_A	N_B	N_C	N_D
Out-of-time region	2.2 ± 0.8	3	8	442	121
In-time region	2.2 ± 0.8	2	8	317	87

5.6.3 Single DT Cluster

The dominant backgrounds in the single-DT-cluster category are punch-through jets and low-p_T particles from pileup. To estimate the background from the low-p_T particles, the ABCD method is used. Like the single-CSC-cluster category, the same discriminating variables $\Delta\phi(\vec{p}_T^{\text{miss}}, \text{cluster})$ and N_{hits} are used, except the thresholds for the signal selection are $\Delta\phi(\vec{p}_T^{\text{miss}}, \text{cluster}) < 1$ and $N_{\text{hits}} > 100$.

The background prediction is validated in a pileup-enriched VR, as shown in Table 5.4. The pileup-enriched region is defined by inverting the loose identification criterion on the leading jet, and removing the RPC hit matching criteria, $\Delta\phi(\vec{p}_T^{\text{miss}}, \text{cluster})$ requirement, and filters that reject non-collision background. To reduce the statistical uncertainties when estimating the background rates in bins C and D, we merge the MB3 and MB4 categories. The final ABCD fit in the SR is also performed with those categories merged.

Table 5.4: Validation of the ABCD method for the single-DT-cluster category in a pileup-enriched region. The uncertainty in the prediction is the statistical uncertainty propagated from bins B, C, and D. Bins C and D for MB3 and MB4 categories are combined to reduce statistical uncertainty in the two regions. The final ABCD fit in the SR will also be performed with those bins combined.

Cluster station	λ_A	N_A	N_B	N_C	N_D
MB2	1.3 ± 0.9	3	2	130	82
MB3	0.6 ± 1.0	1	1	20	11
MB4	0.0 ± 1.1	1	0		

To estimate the punch-through jet background in the SR, which is not accounted for in the ABCD method, we measure the number of observed events in excess of the ABCD prediction in the region with the inner DT station hit veto inverted, and multiply it by $\epsilon/(1 - \epsilon)$ where ϵ is the corresponding veto efficiency. For clusters in MB2, only the MB1 veto is applied, so only the MB1 veto is inverted. For clusters in MB3 and MB4, both MB1 and MB2 vetoes are inverted. The number of excess events in the inverted region is measured to be 22 ± 7, 7 ± 3, and 2.0 ± 1.7 for clusters in MB2, MB3, and MB4, respectively. The veto efficiency ϵ is measured to be 0.23 ± 0.02, 0.38 ± 0.07, 0.29 ± 0.14 for clusters in MB2, MB3, and MB4, respectively. The statistical uncertainties in the measured veto efficiencies and the number of excess events in the inverted region are propagated as a systematic uncertainty on the background prediction.

The inner DT station (MB1 or MB1 plus MB2) hit veto efficiency is measured in a separate punch-through jet enriched region, by selecting clusters that have a jet with $p_T > 10\,\text{GeV}$ within $\Delta R < 0.4$. The measured veto efficiency is measured as a function of the matched jet p_T, as shown in Figure 5.16. The distribution is then fitted with the sum of an exponential and constant function for clusters from each station separately. The veto efficiencies are then extrapolated by evaluating the fit function at 0 GeV to predict the veto efficiencies for clusters passing the jet veto. The punch-through jet background prediction method is validated, as shown in Table 5.5, by predicting background clusters that are matched to 2–5 MB1 or 2–5 MB2 hits, instead of <2 MB1 or <2 MB2 hits in the SR.

Table 5.5: Validation of the punch-through jet background prediction method for the single-DT-cluster category. The uncertainty in the prediction is the statistical uncertainty propagated from the extrapolated MB1/MB2 hit veto efficiency.

Cluster station	λ_A	N_A
MB2	4.7 ± 1.5	6
MB3	1.5 ± 0.9	0
MB4	1.0 ± 0.9	0

5.7 Systematic Uncertainties and Corrections

5.7.1 Background Uncertainties

The only source of systematic uncertainty in the background prediction is the punch-through jet background in the single-DT-cluster category. The size of the uncertainty is 32%, 50%, and 100% for clusters in MB2, MB3, and MB4, respectively. No additional background systematic uncertainties are assigned for the background predicted by the ABCD method because the ABCD background estimation method

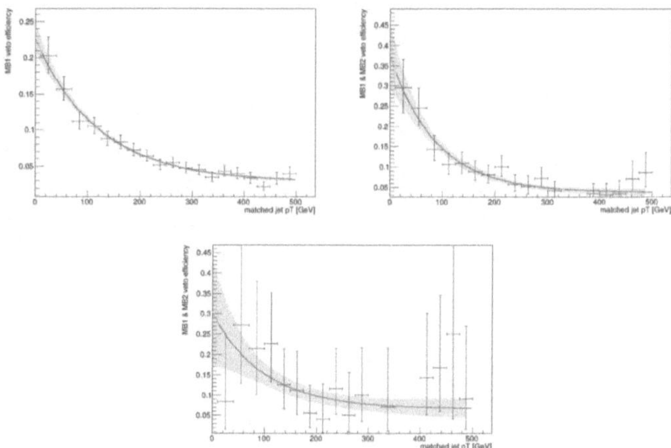

Figure 5.16: The inner DT station (MB1 or MB1 plus MB2) hit veto efficiency for MB2 (top left), MB3 (top right), and MB4 (bottom) clusters measured as a function of matched jet p_T in an inverted jet veto selection. The distributions are fitted to the sum of exponential and constant function for each station separately. The χ^2 per degree of freedom of the fits are 11.1/16, 12.0/16, and 14.2/16 for MB2, MB3, and MB4, respectively.

is validated, as detailed in Section 5.6. Background statistical uncertainty from the ABCD method is propagated to the SR prediction.

5.7.2 Signal Uncertainties

The dominant source of uncertainty in the signal prediction are missing higher-order QCD corrections, which amounts to 21% for the gluon fusion production and a few percent for the other sub-dominant production. The perturbative QCD renormalization and factorization uncertainties are estimated by changing the scales μ_R and μ_F up and down by a factor of 2 from their default values used in the matrix element calculation. The size of the uncertainty is obtained by re-evaluating the signal yield with the different Higgs boson p_T shape with different (μ_R, μ_F) variations. We consider values of $(\mu_R, \mu_F) = (1.0, 2.0)$, $(1.0, 0.5)$, $(2.0, 1.0)$, $(0.5, 1.0)$, $(2.0, 2.0)$, $(0.5, 0.5)$ and re-evaluate the signal yield with the different Higgs boson p_T shape with respect to the nominal signal yield. The total uncertainty is calculated by summing the size of the six variations in quadruture. The Higgs boson p_T shape with different μ_R and μ_F values are shown for the ggH signal in

Figure 5.17. The same procedure is repeated for VBF, VH, and ttH.

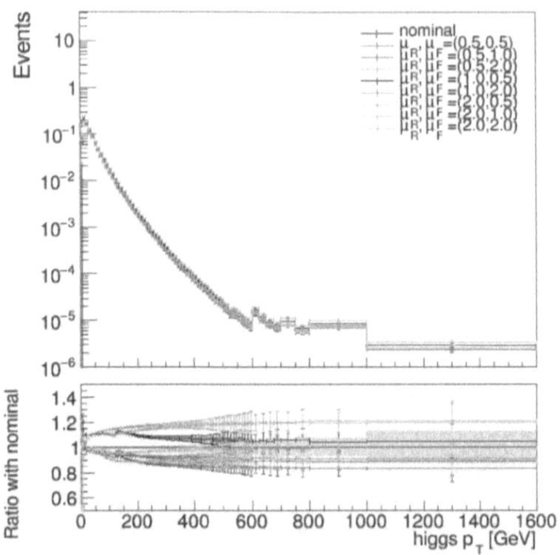

Figure 5.17: The Higgs boson p_T shape calculated from different renormalization and factorization scale.

The uncertainties on the inclusive signal cross section for each production process from QCD scale and PDF variations are taken from the LHC Higgs Cross Section working group yellow report 4 [136] and listed in Table 5.6.

Table 5.6: Higgs boson production cross sections and uncertainties for various production modes at $\sqrt{s} = 13\,\text{TeV}$

Process	Cross secion[pb]	+ QCD scale unc.[%]	- QCD scale unc.[%]	(PDF + α_s) unc. [%]
ggH	48.58	+4.6	-6.7	3.2
VBF	3.782	+0.4	-0.3	2.1
WH	1.373	+0.5	-0.7	1.9
qq$\rightarrow$ ZH	0.884	+0.7	-0.6	1.9
gg$\rightarrow$ ZH	0.123	+25.1	-18.9	2.4
ttH	0.507	+5.8	-9.2	3.6

The other main sources of experimental uncertainty include the signal modeling of the cluster reconstruction and selection criteria, as detailed in the following section, jet energy scale (3–6%) [80], pileup modeling (2%), integrated luminosity (1.6%) [137–139], p_T^{miss} trigger efficiency (5% downward correction and 1% uncertainty), and simulation sample size (3–5%).

5.7.2.1 Cluster Simulation Uncertainties

We studied the simulation modeling of the cluster reconstruction efficiency, cluster-level selections, and veto efficiencies. The accuracy of the simulation prediction for the clusters depends on its modeling of the response of the muon detectors in an environment with multiple particles, each producing many secondary shower particles. This aspect is validated by measuring clusters produced in $Z \rightarrow \mu^+\mu^-$ data events, in which one of the muons undergoes bremsstrahlung in the muon detectors and the associated photon produces an electromagnetic shower. The discrepancy between data and simulation is taken as a correction or systematic uncertainty in the cluster reconstruction and selection efficiencies.

The modeling of the veto efficiencies, including the jet, muon, ME1, MB1, and RPC hit vetoes, is determined from the simulation of jets and muons, the presence of pileup particles, and random noise. The veto efficiencies are measured by randomly sampling the (η, ϕ) locations of clusters from the signal distribution and evaluating whether a jet, muon, or ME1, MB1, or RPC hit from $Z \rightarrow \mu^+\mu^-$ events has been observed within $\Delta R < 0.4$ of the cluster's location. The veto uncertainties are assessed through data and MC comparison, as described below.

For CSC clusters in the single-CSC-cluster category, the systematic uncertainty is dominated by the cluster reconstruction efficiency and the cluster identification efficiency. This uncertainty is determined by measuring the difference in efficiencies in simulation and data, which amounts to an 8% relative uncertainty. Additionally, in signal simulations, the CSC hits are always assumed to be read out, while in the actual data acquisition, only those hits in a chamber that has at least three cathode hits and at least four anode hits at different CSC layers and match predefined hit patterns are read out. This could lead to an underestimation of the efficiency of ME1/1 or ME1/2 vetoes in simulation, which were estimated to have an uncertainty of 1%.

The signal loss from the vetoes is dominated by the muon veto, which is affected by muon segments produced by particles resulting from the LLP decay itself. The simulation modeling of this effect is measured using a control sample of clusters matched to trackless jets made to resemble the signal LLP decay by requiring the neutral energy fraction to be larger than 95%. A 10% downward correction is applied to the signal efficiency to account for the simulation's mis-modeling of the vetoes.

For CSC clusters in the double-cluster category, looser selections are applied, resulting in the systematic uncertainty being dominated by the cluster time spread requirement, which amounts to 10% in the CSC-CSC category and 5% in the DT-CSC category. Furthermore, in the double-cluster category, the jet and muon vetoes are implemented with tighter identification criteria. As a result, the presence of jets and muons is well modeled and no corrections and uncertainty are assigned.

For the DT clusters in the single-DT-cluster category, the systematic uncertainty is dominated by the cluster reconstruction efficiency, which is measured to be 15%. The MB1 and MB2 veto efficiencies are also measured separately by randomly sampling the locations of the clusters from the signal distribution and evaluating whether an MB1 or MB2 hit has been matched. No correction or uncertainty is assigned for the MB2 veto. A 10% downward correction with a 7% uncertainty is applied to the signal efficiency to account for additional noise MB1 hits in data.

For the DT clusters in the double-cluster category, the signal systematic uncertainty is dominated by the cluster reconstruction efficiency, which is measured to be 3% and 1% in the DT-DT and DT-CSC categories, respectively.

5.8 Results and Interpretations

In this section, we report the search results in each category and the interpretation of the combined results of all three categories in the twin Higgs and dark shower models.

5.8.1 Double Clusters

The results of the search in the double-cluster category are shown in Table 5.7. We observed no statistically significant deviation with respect to the SM prediction. The signal and data distributions of N_{clusters} passing the N_{hits} selection are shown in Figure 5.18.

5.8.2 Single CSC Cluster

The corresponding search result in the single-CSC-cluster category is shown in Table 5.8. No statistically significant deviation with respect to the SM prediction is observed. The signal and data distributions of N_{hits} in bins A and D, and $\Delta\phi(\vec{p}_{\text{T}}^{\text{miss}}, \text{cluster})$ in bins A and B are shown in Figure 5.19.

Table 5.7: Number of predicted background and observed events in the double-cluster category. The background prediction is obtained from a fit to the observed data assuming no signal contribution.

	Category	A	B	C	D	BD
	DT-DT	0.06 ± 0.06	—	3.1 ± 1.6	—	0.9 ± 0.7
Background-only fit	CSC-CSC	0.7 ± 0.4	—	4.7 ± 2.0	—	3.6 ± 1.5
	DT-CSC	0.12 ± 0.12	1.9 ± 1.2	14.1 ± 3.8	0.9 ± 0.7	—
	DT-DT	0	—	3	—	1
Observation	CSC-CSC	2	—	6	—	1
	DT-CSC	0	2	14	1	—

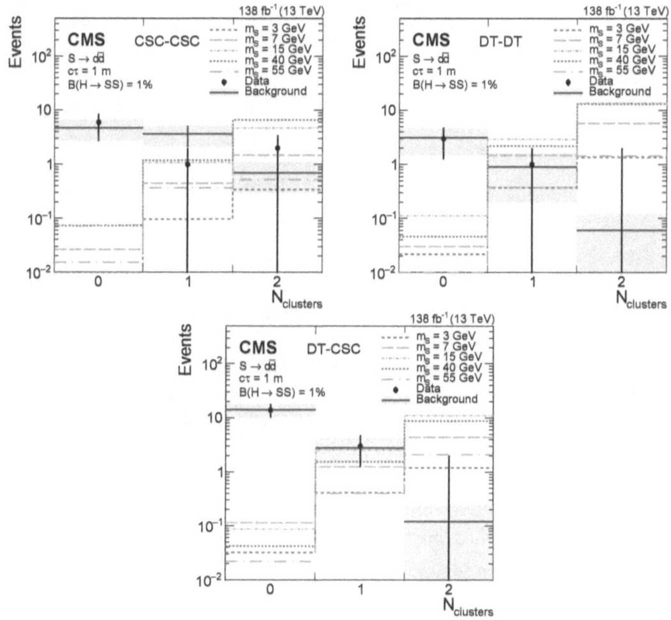

Figure 5.18: The signal (assuming $\mathcal{B}(\mathrm{H} \rightarrow \mathrm{SS}) = 1\%$, $\mathrm{S} \rightarrow \mathrm{d\bar{d}}$, and $c\tau = 1$ m), background, and data distributions of N_{clusters} passing the N_{hits} selection in the search region for CSC-CSC (upper left), DT-DT (upper right), and DT-CSC (lower) categories.

5.8.3 Single DT Cluster

Using the methods detailed in Section 5.6, the background from punch-through jets and low-p_{T} pileup particles are estimated in the single-DT-cluster category. The result of the search is shown in Table 5.9.

Table 5.8: Number of predicted background and observed events in the single-CSC-cluster category. The background prediction is obtained from the fit to the observed data assuming no signal contributions.

	A	B	C	D
Background-only fit	1.8 ± 0.8	4.2 ± 1.7	120 ± 11	51 ± 7
Observed	3	3	121	50

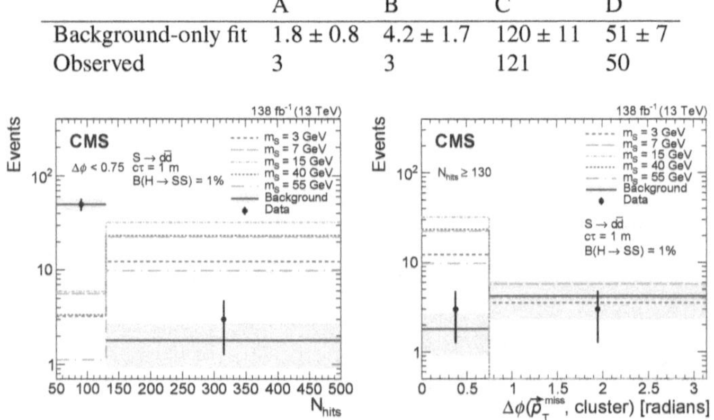

Figure 5.19: Distributions of N_{hits} (left) and $\Delta\phi(\vec{p}_{\text{T}}^{\,\text{miss}}, \text{cluster})$ (right) in the search region of the single-CSC-cluster category. The background predicted by the fit is shown in blue with the shaded region showing the fitted uncertainty. The expected signal with $\mathcal{B}(\text{H} \rightarrow \text{SS}) = 1\%$, $\text{S} \rightarrow \text{d}\bar{\text{d}}$, and $c\tau = 1\,\text{m}$ is shown for m_S of 3, 7, 15, 40, and 55 GeV in various colors and dotted lines. The N_{hits} distribution includes only events in bins A and D, while the $\Delta\phi(\vec{p}_{\text{T}}^{\,\text{miss}}, \text{cluster})$ distribution includes only events in bins A and B. The last bin in the N_{hits} distribution includes overflow events.

We observed no statistically significant deviation with respect to the SM prediction. The signal and data distributions of N_{hits} in bins A and D, and $\Delta\phi(\vec{p}_{\text{T}}^{\,\text{miss}}, \text{cluster})$ in bins A and B are shown in Figure 5.20.

Table 5.9: Number of predicted background and observed events in the single-DT-cluster category. The background prediction is obtained from the fit to the observed data assuming no signal contributions.

	Category	A (total)	A (punch-through)	A (ABCD pred.)	B	C	D
Background-only fit	MB2	9.5 ± 1.9	6.3 ± 1.7	3.1 ± 1.3	4.8 ± 1.9	119.2 ± 11.5	76.8 ± 8.1
	MB3	3.7 ± 1.5	3.1 ± 1.1	0.6 ± 1.1	0.5 ± 0.5	6.5 ± 2.5	7.5 ± 2.6
	MB4	1.2 ± 0.9	1.2 ± 0.9	0.1 ± 0.5	0.06 ± 0.22		
Observation	MB2	9	—	—	5	119	77
	MB3	1	—	—	1	6	8
	MB4	2	—	—	0		

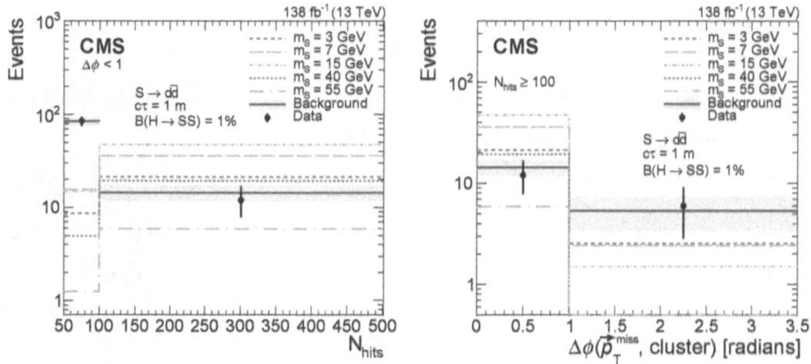

Figure 5.20: Distributions of N_{hits} (left) and $\Delta\phi(\vec{p}_T^{\text{miss}},\text{cluster})$ (right) in the search region of the single-DT-cluster category. The background predicted by the fit is shown in blue with the shaded region showing the fitted uncertainty. The expected signal with $\mathcal{B}(\text{H} \rightarrow \text{SS}) = 1\%$, $\text{S} \rightarrow \text{d}\bar{\text{d}}$, and $c\tau = 1$ m is shown for m_S of 3, 7, 15, 40, and 55 GeV in various colors and dotted lines. The N_{hits} distribution includes only events in bins A and D, while the $\Delta\phi(\vec{p}_T^{\text{miss}},\text{cluster})$ one includes only events in bins A and B. The last bin in the N_{hits} distribution includes overflow events.

5.8.4 Interpretations

In this section, we present the combination of the double-cluster, single-CSC-cluster, and single-DT-cluster categories and the interpretations in the context of the twin Higgs and dark shower models. The ABCD planes of the three categories are defined to be mutually exclusive, and in the combination, we fit the ABCD plane of each category simultaneously.

All theoretical uncertainties assigned to signal simulations are fully correlated. Experimental uncertainties that are not related to the cluster, such as luminosity, jet energy scale, PDFs, and pileup modeling uncertainty, are fully correlated. Experimental uncertainties associated with cluster selections are assumed to be fully uncorrelated. All uncertainties are incorporated into the analysis via nuisance parameters and treated according to the frequentist paradigm.

We evaluate 95% confidence level (CL) upper limits on the branching fractions $\mathcal{B}(\text{H} \rightarrow \text{SS})$ and $\mathcal{B}(\text{H} \rightarrow \Psi\Psi)$ for both the twin Higgs and dark shower models using the modified frequentist criterion CL_s [140, 141] with the profile likelihood ratio test statistic [135]. The Higgs boson production and couplings to the SM particles are assumed as predicted by the SM.

We show the predicted number of signal events for the twin Higgs model in a few benchmark decay, mass, and lifetime scenarios in Table 5.10.

Table 5.10: Expected number of signal events in bin A for each category for a few benchmark signal models assuming $\mathcal{B}(\text{H} \to \text{SS}) = 1\%$.

LLP decay mode, mass, $c\tau$	CSC-CSC	DT-DT	DT-CSC	Single CSC	Single DT
$d\bar{d}$, 3 GeV, $c\tau = 1$ m	0.3	1.3	1.2	12.3	21.2
$d\bar{d}$, 7 GeV, $c\tau = 1$ m	1.5	5.7	4.3	22.5	35.8
$d\bar{d}$, 15 GeV, $c\tau = 1$ m	4.7	13.6	11.1	32.0	46.8
$d\bar{d}$, 40 GeV, $c\tau = 1$ m	6.6	12.9	8.8	23.4	19.3
$d\bar{d}$, 55 GeV, $c\tau = 1$ m	0.5	1.4	2.1	9.8	5.9
$\tau^+\tau^-$, 7 GeV, $c\tau = 1$ m	0.6	1.8	1.6	14.2	22.5
$\tau^+\tau^-$, 15 GeV, $c\tau = 1$ m	1.7	5.2	3.9	20.1	28.9
$\tau^+\tau^-$, 40 GeV, $c\tau = 1$ m	3.3	4.5	3.3	21.3	17.0
$\tau^+\tau^-$, 55 GeV, $c\tau = 1$ m	0.3	0.9	1.0	10.6	6.0
$\pi^0\pi^0$, 0.4 GeV, $c\tau = 0.1$ m	0.1	0.4	0.4	6.8	19.2
$\pi^0\pi^0$, 1 GeV, $c\tau = 0.1$ m	0.4	1.3	1.1	11.6	30.7

The upper limits for the twin Higgs model are shown in Figure 5.21 for all decay modes considered as functions of $c\tau$ for a selection of m_S values. The decay modes include $\text{S} \to d\bar{d}$, $\text{S} \to b\bar{b}$, $\text{S} \to \tau^+\tau^-$, $\text{S} \to \pi^0\pi^0$, $\text{S} \to \pi^+\pi^-$, $\text{S} \to \text{KK}$, $\text{S} \to b\bar{b}$, $\text{S} \to e^+e^-$, and $\text{S} \to \gamma\gamma$.

The sensitivity is mass independent for all decay modes that are shown, except for 55 GeV. The sensitivity for 55 GeV is slightly lower than that of the other mass points because the two LLPs in the event are geometrically closer to each other at higher masses, so when both decay in the muon detectors, they are likely to merge to create one muon detector shower, instead of two distinct clusters, lowering the sensitivity of the double-cluster category. Due to the differences in geometric acceptance, the sensitivity at shorter proper decay lengths is dominated by the single-CSC-cluster category, at longer proper decay lengths by the single-DT-cluster category. At intermediate proper decay lengths, where the analysis is most sensitive, the sensitivity is dominated by the double-cluster category for fully hadronic decays of the LLPs. The addition of the single-DT and double-cluster categories improves upon the previous result based on only CSC clusters [1] by a factor of 2 for $c\tau$ above 0.2, 0.5, 2.0, and 18.0 m for LLP masses 7, 15, 40, and 55 GeV, respectively. The limits for electromagnetic decay modes are slightly less stringent, because the efficiency for reconstructing the corresponding MDS object is smaller.

The upper limits as functions of both mass and $c\tau$ for the $\text{S} \to d\bar{d}$, $\text{S} \to b\bar{b}$, $\text{S} \to \tau^+\tau^-$, $\text{S} \to \pi^0\pi^0$, $\text{S} \to \pi^+\pi^-$, $\text{S} \to \text{KK}$, $\text{S} \to b\bar{b}$, $\text{S} \to e^+e^-$, and $\text{S} \to \gamma\gamma$

are shown in Figure 5.22 and extend as low as $m_S = 0.4\,\text{GeV}$. The analysis is expected to be sensitive below $0.4\,\text{GeV}$ as well because of the calorimetric nature of the signature. However, because of limitations in the signal generation, the analysis sensitivity below $0.4\,\text{GeV}$ is not quantified.

We also consider an explicit scenario where the branching fractions of the S decays are identical to those of a Higgs boson evaluated at m_S [142]. Figure 5.23 shows the upper limits as functions of both mass and $c\tau$ using the branching fractions for S calculated in Ref. [142]. This search provides the most stringent constraint on $\mathcal{B}(H \rightarrow SS)$ for proper decay lengths in the range of 0.04–0.40 m and above 4 m, of 0.3–0.9 m and above 3 m, and of above 0.8 m for an LLP mass of 15, 40, and 55 GeV, respectively. For LLP masses below 10 GeV, this search provides the most stringent constraints on LLPs decaying to particles other than muons.

The observed upper limits for the dark shower models are shown in Figure 5.25–5.28 for the vector, gluon, photon, higgs, and dark photon portals, as a function of $c\tau$ for a selection of m_Ψ, $x_{i\omega}$, and $x_{i\Lambda}$ values. For each portal, we show the upper limits assuming $(x_{i\omega}, x_{i\Lambda}) = (1.0, 1.0)$, $(2.5, 1.0)$, and $(2.5, 2.5)$ and for different LLP masses ranging from 2 to 20 GeV. This search sets the first LHC limits on models of dark showers produced via Higgs boson decay, and is sensitive to the branching fraction of the Higgs boson to dark quarks as low as 2×10^{-3}.

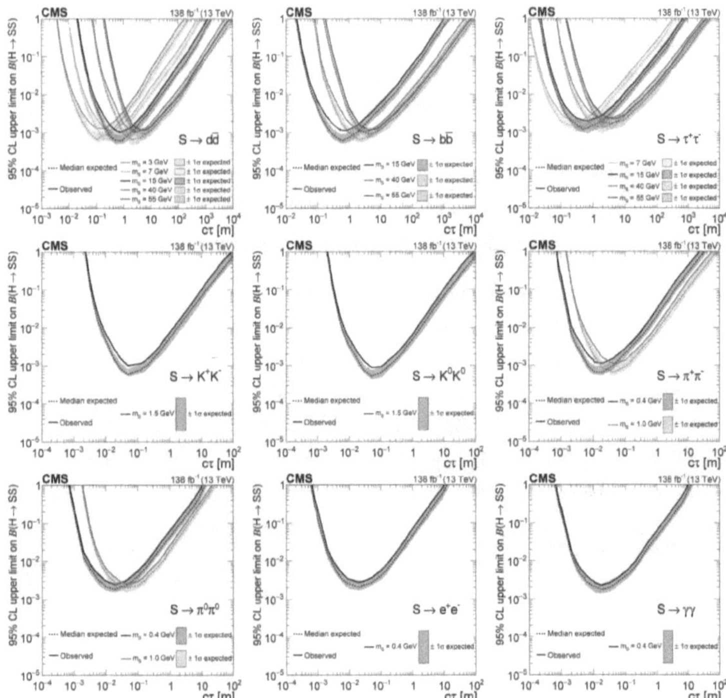

Figure 5.21: The 95% CL expected (dotted curves) and observed (solid curves) upper limits on the branching fraction $\mathcal{B}(\mathrm{H} \to \mathrm{SS})$ as functions of $c\tau$ for the $\mathrm{S} \to \mathrm{d\bar{d}}$ (upper left), $\mathrm{S} \to \mathrm{b\bar{b}}$ (upper center), $\mathrm{S} \to \tau^+\tau^-$ (upper right), $\mathrm{S} \to \mathrm{K}^+\mathrm{K}^-$ (middle left), $\mathrm{S} \to \mathrm{K}^0\mathrm{K}^0$ (middle center), $\mathrm{S} \to \pi^+\pi^-$ (middle right), $\mathrm{S} \to \pi^0\pi^0$ (bottom left), $\mathrm{S} \to \mathrm{e}^+\mathrm{e}^-$ (bottom center), and $\mathrm{S} \to \gamma\gamma$ (bottom right) decay modes. The exclusion limits are shown for different mass hypotheses.

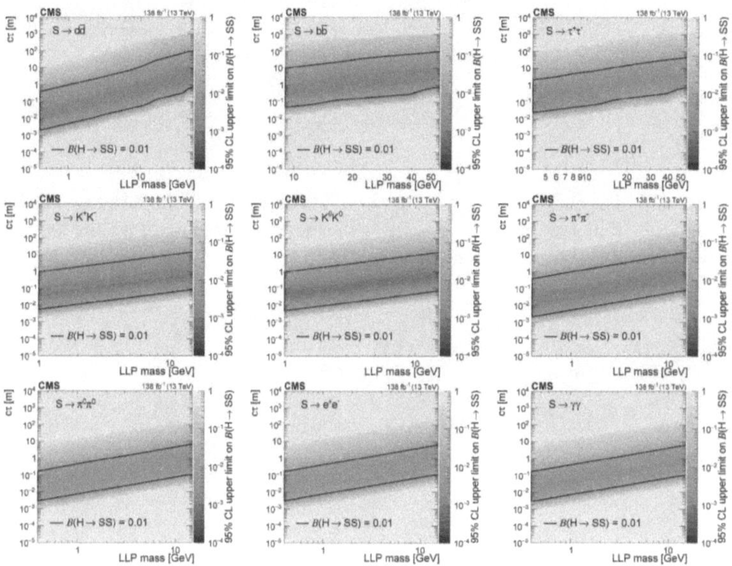

Figure 5.22: The 95% CL observed upper limits on the branching fraction $\mathcal{B}(H \rightarrow SS)$ as functions of mass and $c\tau$ for the $S \rightarrow d\bar{d}$ (upper left), $S \rightarrow b\bar{b}$ (upper center), and $S \rightarrow \tau^+\tau^-$ (upper right) $S \rightarrow K^+K^-$ (middle left), $S \rightarrow K^0K^0$ (middle center), and $S \rightarrow \pi^+\pi^-$ (middle right) $S \rightarrow \pi^0\pi^0$ (bottom left), $S \rightarrow e^+e^-$ (bottom center), and $S \rightarrow \gamma\gamma$ (bottom right) decay modes.

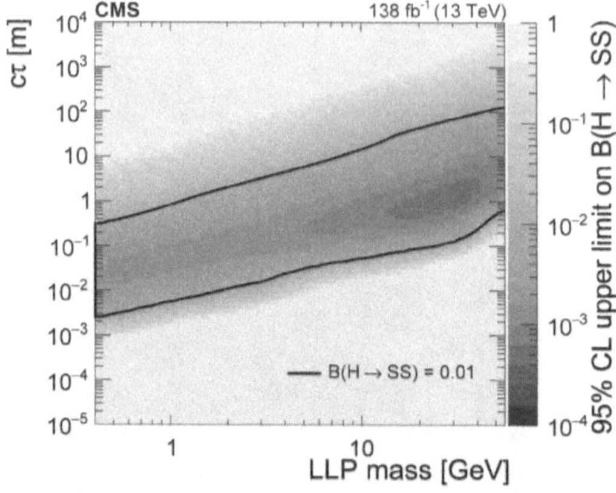

Figure 5.23: The 95% CL observed upper limits on the branching fraction $\mathcal{B}(\mathrm{H} \to \mathrm{SS})$ as a function of mass and $c\tau$, assuming the branching fractions for S are identical to those of a Higgs boson evaluated at m_S [142].

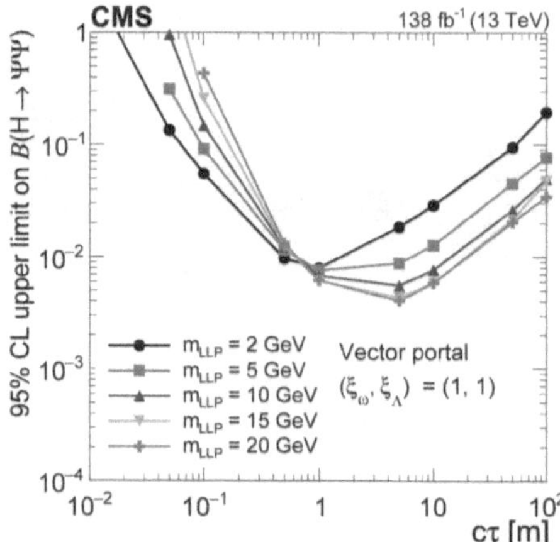

Figure 5.24: The 95% CL observed upper limits on the branching fraction $\mathcal{B}(\mathrm{H} \to \Psi\Psi)$ as functions of $c\tau$ for the vector portal assuming $(x_{i\omega}, x_{i\Lambda}) = (1, 1)$. For the vector portal, scenarios with $x_{i\omega} = 2.5$ are not interpreted, because in this case all vector mesons would decay to scalar mesons that are invisible to the detector. The exclusion limits are shown for different LLP mass hypotheses. The limits are calculated only at the proper decay lengths indicated by the markers and the lines connecting the markers are linear interpolations.

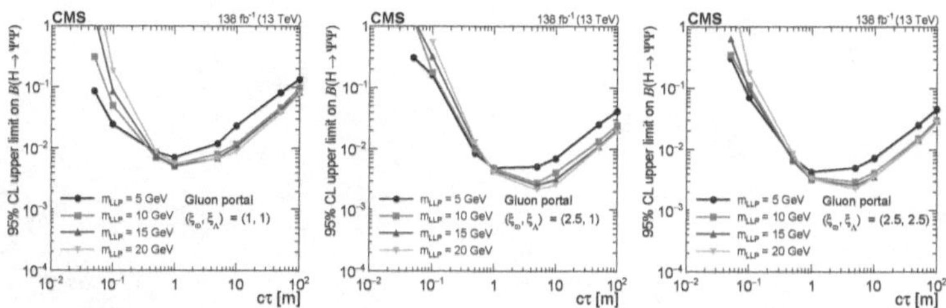

Figure 5.25: The 95% CL observed upper limits on the branching fraction $\mathcal{B}(\mathrm{H} \to \Psi\Psi)$ as functions of $c\tau$ for the gluon portal, assuming $(x_{i\omega}, x_{i\Lambda}) = (1, 1)$ (left), $(x_{i\omega}, x_{i\Lambda}) = (2.5, 1)$ (middle), and $(x_{i\omega}, x_{i\Lambda}) = (2.5, 2.5)$ (right). The exclusion limits are shown for different LLP mass hypotheses. The limits are calculated only at the proper decay lengths indicated by the markers and the lines connecting the markers are linear interpolations.

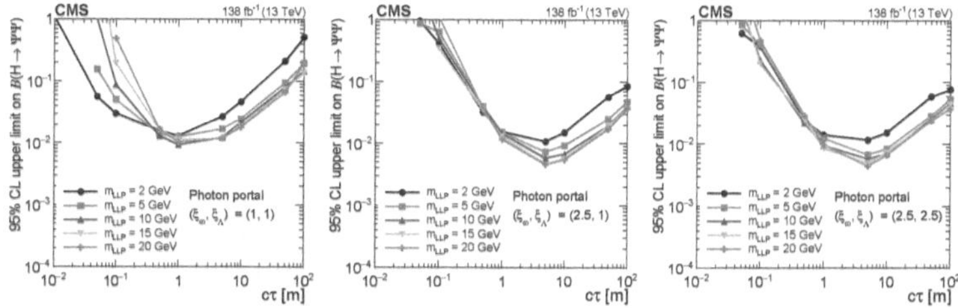

Figure 5.26: The 95% CL observed upper limits on the branching fraction $\mathcal{B}(\mathrm{H} \to \Psi\Psi)$ as functions of $c\tau$ for the photon portal, assuming $(x_{i\omega}, x_{i\Lambda}) = (1, 1)$ (left), $(x_{i\omega}, x_{i\Lambda}) = (2.5, 1)$ (middle), and $(x_{i\omega}, x_{i\Lambda}) = (2.5, 2.5)$ (right). The exclusion limits are shown for different LLP mass hypotheses. The limits are calculated only at the proper decay lengths indicated by the markers and the lines connecting the markers are linear interpolations.

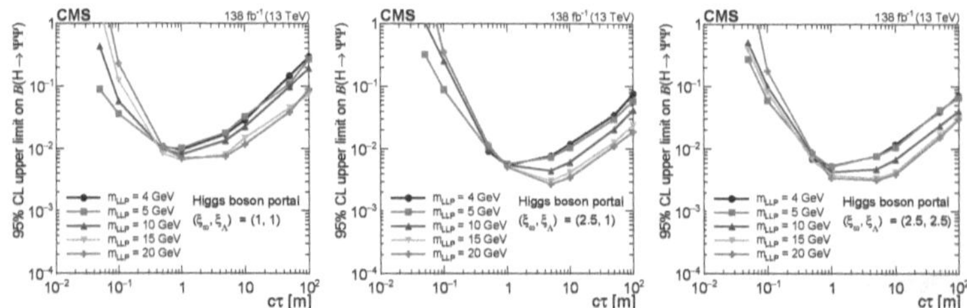

Figure 5.27: The 95% CL observed upper limits on the branching fraction $\mathcal{B}(H \to \Psi\Psi)$ as functions of $c\tau$ for the Higgs boson portal, assuming $(x_{i\omega}, x_{i\Lambda}) = (1, 1)$ (left), $(x_{i\omega}, x_{i\Lambda}) = (2.5, 1)$ (middle), and $(x_{i\omega}, x_{i\Lambda}) = (2.5, 2.5)$ (right). The exclusion limits are shown for different LLP mass hypotheses. The limits are calculated only at the proper decay lengths indicated by the markers and the lines connecting the markers are linear interpolations.

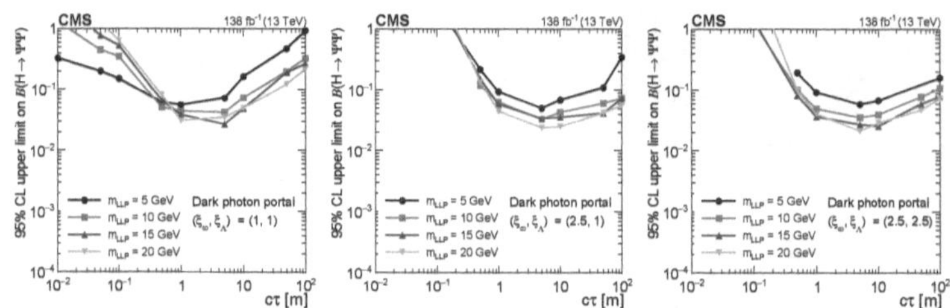

Figure 5.28: The 95% CL observed upper limits on the branching fraction $\mathcal{B}(H \to \Psi\Psi)$ as functions of $c\tau$ for the dark-photon portal, assuming $(x_{i\omega}, x_{i\Lambda}) = (1, 1)$ (left), $(x_{i\omega}, x_{i\Lambda}) = (2.5, 1)$ (middle), and $(x_{i\omega}, x_{i\Lambda}) = (2.5, 2.5)$ (right). The exclusion limits are shown for different LLP mass hypotheses. The limits are calculated only at the proper decay lengths indicated by the markers and the lines connecting the markers are linear interpolations.

5.9 Summary

In summary, data from proton-proton collision at $\sqrt{s} = 13$ TeV recorded by the CMS experiment in 2016–2018, corresponding to an integrated luminosity of $138\,\mathrm{fb}^{-1}$, have been used to conduct the first search that uses both the barrel and endcap CMS muon detectors to detect hadronic and electromagnetic showers from long-lived particle (LLP) decays. Based on this unique detector signature, the search is largely model-independent, with sensitivity to a broad range of LLP decay modes and masses below the GeV scale. With the excellent shielding provided by the inner CMS detector, the CMS magnet and its steel flux-return yoke, the background is suppressed to a low level and a search for both single and pairs of LLPs decays is possible.

No significant deviation from the standard model background is observed. The most stringent LHC constraints to date are set on the branching fraction of the Higgs boson (H) to LLPs with masses below 10 GeV and decaying to particles other than muons. For higher LLP masses the search provides the most stringent branching fraction limits for specific proper decay lengths: 0.04–0.40 m and above 5 m for 15 GeV LLP; 0.3–0.9 m and above 3 m for 40 GeV LLP; and above 0.9 m for 55 GeV LLP. Finally, the first LHC limits on models of dark showers produced via H decay are set, and constrain branching fractions of the H decay to dark quarks as low as 2×10^{-3} at 95% confidence level.

5.A Background Muon Detector Shower from Low p_T Particles

In this section, we present a study for the background composition of the clusters passing all the vetos in the single CSC cluster category in MC samples.

The samples that are being used are W + jet, minimum bias events, and privately produced particle guns consisting of K_L^0, K^+, and π^+ each with p_T of 2, 5, and 10 GeV, respectively.

Due to the limited availability of background simulation samples with the CSC and DT rechit collections available, the only signal region sample available for the background composition study was the W + jet sample. Private reprocessing of this sample was necessary to gain access to the CSC and DT rechit collections because the AODSIM data-tier of the centrally produced samples does not contain these rechit collections. The CSC and DT rechit clusters from this W + jet sample are representative of the clusters in the final signal region. There are a total of 7.1×10^7 events in the sample, and after we apply most of the cluster-level selections, including the jet, muon, ME11/12, RE12, RB1, MB1 segment veto, time spread cut, time cut, and N_{hits} cut, we are left with 1084 events with one CSC cluster.

We found that among the 1084 events, 34% of the clusters are matched to a generator-level muon. These clusters were all required to pass the muon veto, and therefore are cases where the muon reconstruction failed. The overwhelming majority of such clusters are matched to generator-level muons with large η values near the edge of the CSC acceptance, as shown in Figure 5.29, and for this reason often fail to be reconstructed. In our analysis, we apply the $|\eta| < 2$ cut to explicitly reject such background clusters.

Of the remaining clusters, we observed that the majority of clusters are matched to low p_T generator-level kaons and pions, as shown in Table 5.11. Initially, we suspected that the matching between these clusters and the low p_T pions and kaons were accidental and that other particles from pileup may be the real cause for the cluster objects, and were not found simply because generator-level pileup particles are not saved in the event record.

To verify this hypothesis, we performed the same study using the minimum bias simulation sample without any pileup events mixed in. The minimum bias sample has a total of 3.3×10^8 events, and after we apply the same cluster-level selections, we found 76 events with one CSC cluster, which results in a cluster efficiency of about 2.3e-7. Similarly, we match the CSC clusters to status 1 generator-level particles to

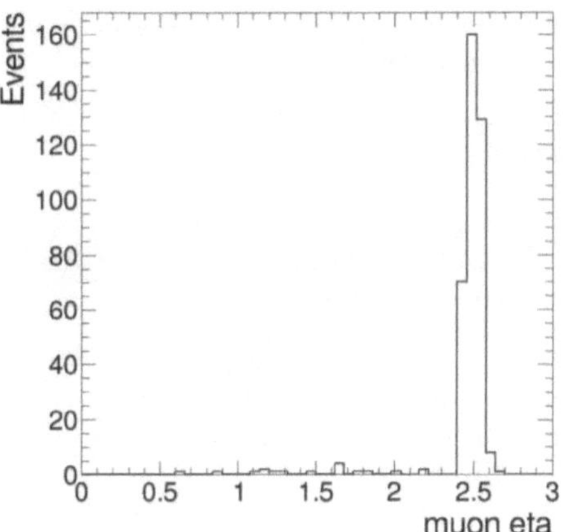

Figure 5.29: The distribution of generator-level muon $|\eta|$ of the muons that are matched to clusters passing all vetos in the W + jet sample.

Table 5.11: A breakdown of the generator-level particles that are matched to background CSC clusters passing the cluster selections in the W + jet sample. The cases where no match was found are interpreted to be caused by pileup particles, for which generator-level particles are not stored in the event record.

particle type	fraction of events
K^0_L	3.6%
K^0_S	3.1%
charged kaon $p_T > 2$ GeV	0.6%
charged pion $p_T > 2$ GeV	2.5%
charged kaon $1.5 < p_T < 2.5$ GeV	1.1%
charged pion $1.5 < p_T < 2.5$ GeV	5.2%
charged kaon $0.5 < p_T < 1.5$ GeV	22.9%
charged pion $0.5 < p_T < 1.5$ GeV	44.2%
not matched to any status 1 genParticles	16.8%

study the origin of the background clusters. We found that similar to the W + jet sample after removing the muon contribution, most clusters are matched to low p_T pions and kaons, as shown in Table 5.12.

To confirm that the background clusters really are produced by these low p_T pions and kaons, suggested by the W + jet and minimum bias simulation samples, we produced particle gun samples for K^0_L, K^+, and π^+ with p_T of 2, 5, and 10 GeV

Table 5.12: Breakdown of the type of status 1 generator-level particles that CSC clusters are matched to in minimum bias sample.

particle type	fraction of events
muon	3.9%
K_L^0	7.9%
K_S^0	5.3%
charged kaon $p_T > 2$ GeV	2.6%
charged pion $p_T > 2$ GeV	5.3%
charged kaon $1.5 < p_T < 2.5$ GeV	2.6%
charged pion $1.5 < p_T < 2.5$ GeV	5.3%
charged kaon $0.5 < p_T < 1.5$ GeV	23.7%
charged pion $0.5 < p_T < 1.5$ GeV	43.4%

in order to verify that such low momentum single particles can in fact produce the CSC clusters of the type observed in our background samples.

We observed that the efficiency of reconstructing a cluster (without any vetos) is of the order of 10^{-5} to 10^{-3} for both kaons and pions, increasing with p_T of the particle. We further observed that the efficiency of reconstructing a cluster passing all selections is on the order of 10^{-6} to 10^{-7}, which roughly agree with what we observed in the minimum bias simulation sample. A summary of the particle gun cluster efficiencies are shown in Table 5.13.

Table 5.13: Summary of cluster efficiencies for single particle guns. We study K_L^0, K^+, and π^+ with p_T ranging from 2 to 10 GeV.

sample, p_T	Total N_{events}	cluster efficiency (No veto)	efficiency with all selections (except for $N_{\text{hits}} > 130$)	efficiency with all selections including $N_{\text{hits}} > 130$
K_L^0, 2 GeV	2.5E+07	2.4E-05	2.8E-07	0E+00
K_L^0, 5 GeV	2.5E+07	2.4E-04	1.9E-06	1.6E-07
K_L^0, 10 GeV	2.4E+07	9.3E-04	5.3E-06	1.8E-06
K^+, 2 GeV	2.5E+07	2.2E-05	3.7E-07	0E+00
K^+, 5 GeV	2.5E+07	3.4E-04	1.5E-06	4.0E-08
K^+, 10 GeV	2.3E+07	9.7E-04	1.8E-06	1.3E-07
π^+, 2 GeV	2.5E+07	1.7E-05	4.1E-08	0E+00
π^+, 5 GeV	2.5E+07	3.3E-04	7.3E-07	0E+00
π^+, 10 GeV	2.3E+07	1.4E-03	1.3E-06	1.3E-07

We also show in Figure 5.30 the N_{hits} distribution for the clusters from 2 GeV K_L^0,

K^+, and π^+, demonstrating that, although very rarely, low p_T pions and kaons can produce clusters with a large number of rechits.

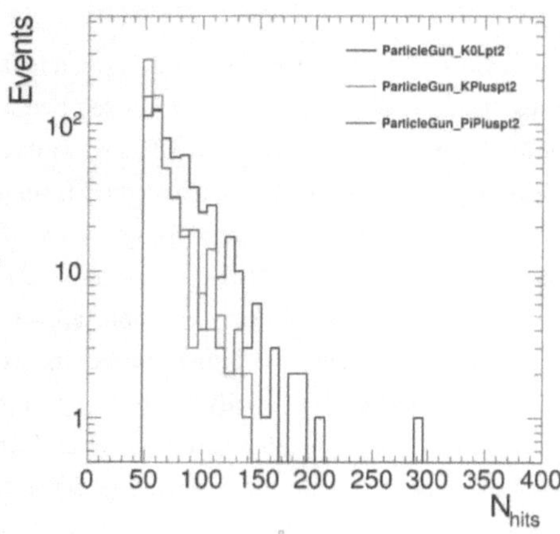

Figure 5.30: The distribution of N_{hits} for the clusters in the 2 GeV particle gun samples.

Based on these studies with W + jet, minimum bias, and particle gun simulation samples, we conclude that after all the cluster-level vetos that reject clusters from jets and muons with high p_T, the dominant background clusters are from low p_T pions and kaons. They will very rarely punchthrough the shielding material and produce background clusters with large cluster size. However, due to the large flux of low p_T pions and kaons in LHC collision events, they are the dominant source of background CSC clusters.

Re-interpretation of LLPs Decaying in the CMS Endcap Muon Detectors

6.1 Introduction

The search for LLPs using CMS muon detectors, as described in the previous chapter, presented the results on a benchmark model motivated by the twin Higgs and dark shower scenario. To further extend the physics reach of this novel shower-like signature, the model independence of the MDS in endcap was studied in the context of the first paper [1], and ongoing studies are in progress for MDS in the barrel region. The reconstruction efficiency of MDS can be parameterized as a function of generator-level LLP energy and LLP decay position, allowing for reinterpretation of the search in any theoretical models that predict the existence of LLPs. The parameterized cluster efficiency was published in the HEPData entry [143] of the endcap-only analysis as a function of generator-level LLP information. The published parameterized functions were implemented in dedicated new DELPHES modules. Using the new DELPHES module, I have reinterpreted the endcap-only search in a number of models, including the dark scalar, dark photon, axion-like particles, inelastic dark matter, hidden valley models, and heavy neutral leptons.

This chapter is structured as follows. In Section 6.2, the study of the parameterized efficiency for MDS, which is published in the HEPData entry [143] of the first endcap-only analysis, is presented. Section 6.3 outlines the generation and simulation framework, the validation of the framework against the CMS result, and analysis strategy. All benchmark models that are considered are introduced in Section 6.4. Finally, in Section 6.5, the recasts and sensitivity projections of this signature in all of the benchmark models are shown. The results discussed in this chapter are published in two papers [3, 4].

This chapter contains published work from [1], [3], and [4]. Text from Section 6.2 and Figures 6.2, 6.5, and 6.6 are adapted with permission from CMS Collaboration. "Search for long-lived particles decaying in the CMS end cap muon detectors in proton-proton collisions at $\sqrt{s}$ =13 TeV.". In: *Physical Review Letters* 127.26 (2021), p. 261804. DOI: 10.1103/PhysRevLett.127.261804. arXiv: 2107.04838 [hep-ex]. Copyright © 2021 American Physical Society.

Text from Sections 6.3–6.6 and Figures 6.8–6.22 are adapted with permission from Andrea Mitridate et al. "Energetic long-lived particles in the CMS muon chambers." In: *Physical Review D* 108.5 (2023), p. 055040. DOI: 10.1103/PhysRevD.108.055040. arXiv: 2304.06109 [hep-ph]. © 2023 American Physical Society and Giovanna Cottin et al. "Long-lived heavy neutral leptons with a displaced shower signature at CMS.". In: *Journal of High Energy Physics* 02 (2023), p. 011. DOI: 10.1007/JHEP02(2023)011. arXiv: 2210.17446 [hep-ph]. © 2023 Springer Nature.

6.2 Parameterized Efficiency for Muon Detector Showers

This section documents the signal efficiency parameterization that was published in the HEPData entry [143] of the endcap only analysis [1]. Parameterizations for both cluster efficiency that includes all cluster-level selections (excluding jet veto, time, and $\Delta\phi(\vec{p}_T^{\text{miss}}, \text{cluster})$ cut) and the cut-based ID efficiency per LLP were released, allowing for reinterpretation of the analysis in any models with LLPs. The jet veto, time, and $\Delta\phi(\vec{p}_T^{\text{miss}}, \text{cluster})$ requirements are model-dependent and can be implemented from the generator-level information, as detailed in Section 6.3. A simple parameterization as a function of generator-level LLP hadronic energy, electromagnetic (EM) energy, and decay position is sufficient to reproduce full simulation signal efficiencies for LLPs to within 35% and 20% in the geometric acceptance region A and B, as defined below, respectively. The study is validated for LLPs with mass between 7-55 GeV, lifetime between 0.1 m–100 m, and $S \rightarrow d\bar{d}$ and $S \rightarrow \tau^+\tau^-$ decay modes.

The electromagnetic and hadronic energy of the LLP is calculated by first matching status 1 (final state) generator-level particles to the LLP if the particle production vertex is within 0.1 m from the LLP decay vertex. If the particle is matched to both LLPs, then it is assigned to the closer LLP. The energy of the matched generator-level particle is assigned as EM energy if the particle is a neutral pion, electron, or photon. The energy of the matched particle is ignored if it's a neutrino or muon, as they do not produce showers in the muon system. All other particles are assigned as hadronic energy. The parameterization efficiency is derived by using the $S \rightarrow \tau^+\tau^-$ signal simulation samples, since τ leptons decay both leptonically and hadronically.

The LLP decay location is categorized into two regions, as shown in Figure 6.1. These two regions have qualitatively different behaviors, so the parameterized efficiency is derived for each region separately. Within each region, they have quantitatively similar behavior. Region A is defined as 391 cm $< r <$ 695.5 cm and 400 cm $< |z| <$ 671 cm. Region B is defined as 671 cm $< |z| <$ 1100 cm, $r <$ 695.5 cm and $|\eta| < 2$.

The fraction of LLPs that decay in each region are dependent on the LLP mass and $c\tau$. However, the signal efficiency in A is much lower than that in B due to the ME11/12 veto, so more than 90% of clusters in the signal region are from LLPs that decay in region B.

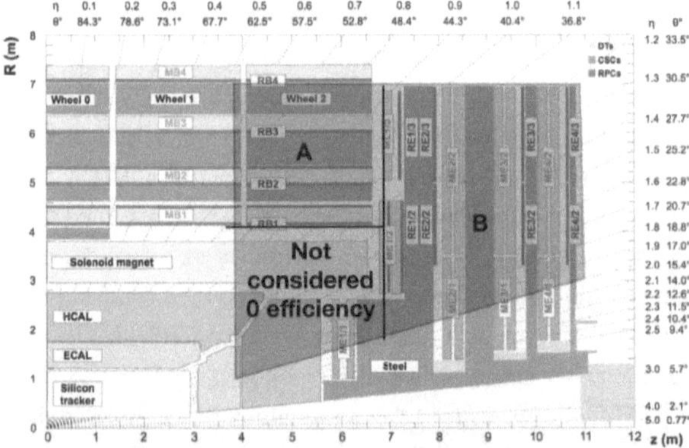

Figure 6.1: The geometric acceptance region considered for LLP decay in the endcap is shaded in red. Region A and B are shown. Region A is defined as 391 cm $< r <$ 695.5 cm and 400 cm $< |z| <$ 671 cm. Region B is defined as 671 cm $< |z| <$ 1100 cm and $r <$ 695.5 cm and $|\eta| <$ 2. The rest of the acceptance region are not considered, since the signal efficiency is almost zero ($<$0.5%), due to shielding and the vetos.

6.2.1 Cluster Efficiency

The cluster efficiency is measured in the two decay regions separately in bins of hadronic and EM energy, as shown in Figure 6.2. The cluster efficiency is defined by the cluster reconstruction efficiency multiplied by the signal efficiency of the cluster-level selections, including the muon veto, segment/rechit vetos, time spread selection, and $N_{\text{hits}} >$ 130 per LLP. The cluster efficiency parameterization does not include the jet veto, time, and $\Delta\phi(\vec{p}_{\text{T}}^{\text{miss}}, \text{cluster})$ cut, since these cuts are model dependent and can be easily implemented using generator-level information.

The parameterization procedure is validated to be independent of LLP mass (7-55 GeV) and lifetime (0.1 m–100 m) using the S $\rightarrow \tau^+\tau^-$ signal sample. Since the EM fraction does not change with LLP mass and lifetime, the cluster efficiency with respect to the LLP energy is shown for different LLP masses and different lifetimes to allow for easier visualization, as shown in in Figure 6.3 and Figure 6.4, respectively. No dependence on LLP mass (7–55 GeV) and lifetime (0.1 m–100 m) are observed.

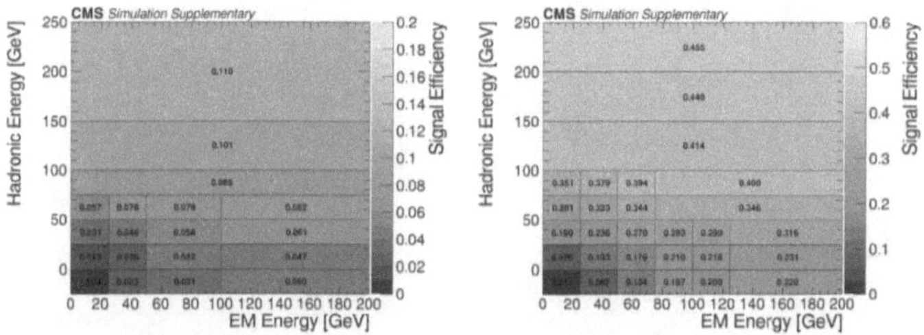

Figure 6.2: The cluster efficiency in bins of hadronic and EM energy in region A (left) and B (right). The cluster efficiency is evaluated using the sum of all mass and $c\tau$ models available from the 4τ sample. The first hadronic energy bins correspond to LLPs that decayed leptonically with 0 hadronic energy. The statistical uncertainty for each bin is documented in Additional Figure 7 of the HEPData record of this analysis [143].

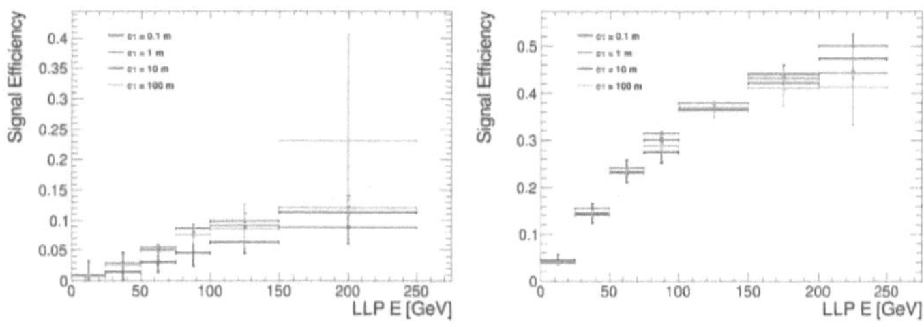

Figure 6.3: The cluster efficiency with respect to LLP energy in region A (left) and B (right) for a 15 GeV LLP and different LLP lifetimes. The plots demonstrate the cluster efficiency is independent of the LLP lifetime.

The independence of the LLP decay mode is also validated by checking the cluster efficiency in bins of hadronic and EM energy for the $S \rightarrow d\bar{d}$ signal. The parameterization derived from $S \rightarrow d\bar{d}$ signal, as shown in Figure 6.5 is in good agreement with the cluster efficiency derived from the $S \rightarrow \tau^+\tau^-$ signal, shown in Figure 6.2,

Finally, the signal yield prediction is validated against the full simulation prediction for models with varying LLP mass between 7-55 GeV, lifetime between 0.1–100 m, and decay mode $d\bar{d}$ and $\tau^-\tau^+$, using the parameterization shown in Figure 6.2. The parametrized signal yield for all models agrees with the full simulation prediction to within 35% and 20% for region A and B, respectively.

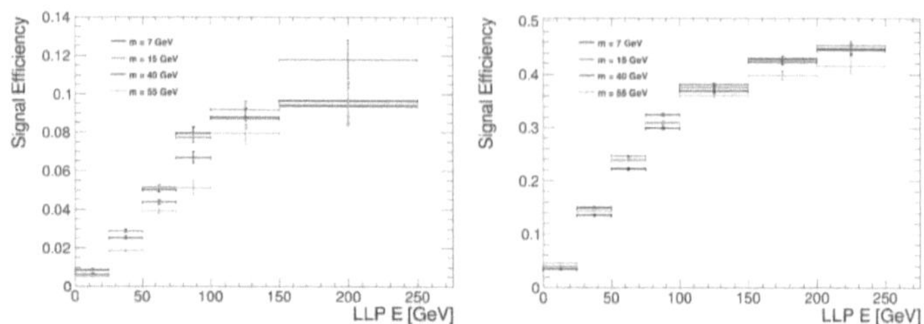

Figure 6.4: The cluster efficiency with respect to LLP energy in region A (left) and B (right) for different LLP masses. The different LLP mass samples consists of the sum of all available lifetimes ranging from 0.1 to 100 m to improve the statistics. The plots demonstrate the cluster efficiency is independent of the LLP mass.

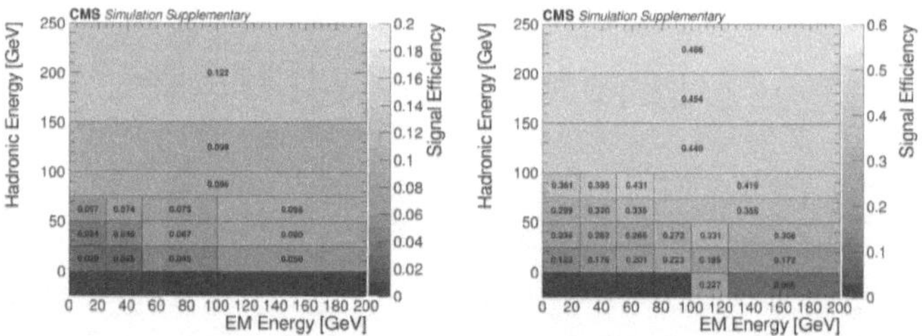

Figure 6.5: The cluster efficiency estimated from LLP decaying to dd in bins of hadronic and EM energy in region A (left) and B (right). The first hadronic energy bins correspond to LLPs that decayed leptonically with 0 hadronic energy, so it's empty for d$\bar{\text{d}}$ decays. The sample includes the sum of all available mass (7–55 GeV) and ctau (0.1–100 m) points. The parameterization agrees with that derived from the $S \rightarrow \tau^+\tau^-$ signal.

6.2.2 Cut-Based ID Efficiency

The cut-based ID efficiency is parameterized with respect to clusters that pass the cluster-level selections. As described in Section 5.5, the cluster ID requirement applies different η cuts depending on the N_{stations} and average station number of the cluster.

Since the entire region A has lower η than the most stringent η cuts, all clusters in region A pass the cut-based ID. Therefore, we focus on the parameterization of cut-based ID efficiency in region B only.

Two steps are needed to parametrize the efficiency of the cut-based ID. We need a parameterization of the efficiency of $N_{\text{stations}} > 1$ requirement and a transfer function that takes generator-level LLP decay position as input and outputs the reconstructed cluster average station (only for clusters with $N_{\text{stations}} = 1$)

The efficiency of $N_{\text{stations}} > 1$ can be well parameterized using just the generator-level hadronic energy in the region B, as shown in Figure 6.6. It has been validated that the parameterization is independent of LLP mass (7–55 GeV), lifetime (0.1–100 m), and decay mode($d\bar{d}$ and $\tau^+\tau^-$) within region B, as shown in Figure 6.7.

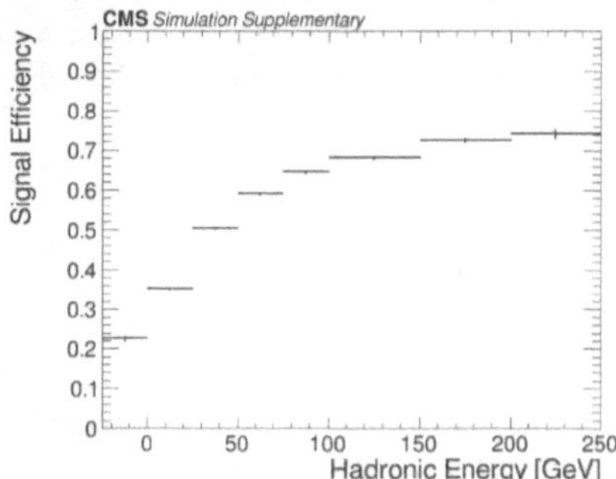

Figure 6.6: The efficiency of $N_{\text{stations}} > 1$ with respect to hadronic energy in region B. The first hadronic energy bin corresponds to LLPs that decayed leptonically with 0 hadronic energy.

The average station transfer function is implemented with a simple mapping between the LLP decay position and the average station number. The cluster station is defined to be the station subsequent to the LLP decay position. For example, if the LLP decays between station 2 and 3, the average station will be station 3. The average station function and an implementation of cut-based ID is provided as part of the *Additional Resources* in the HEPData entry [143].

With the $N_{\text{stations}} > 1$ efficiency parameterization and transfer function in place, the signal yield prediction was validated against the full simulation prediction for models with varying LLP masses between 7-55 GeV, lifetimes between 0.1–100 m, and decay modes $d\bar{d}$ and $\tau^-\tau^+$. The parametrized signal yield matches the full simulation prediction to within 10% for all signal models.

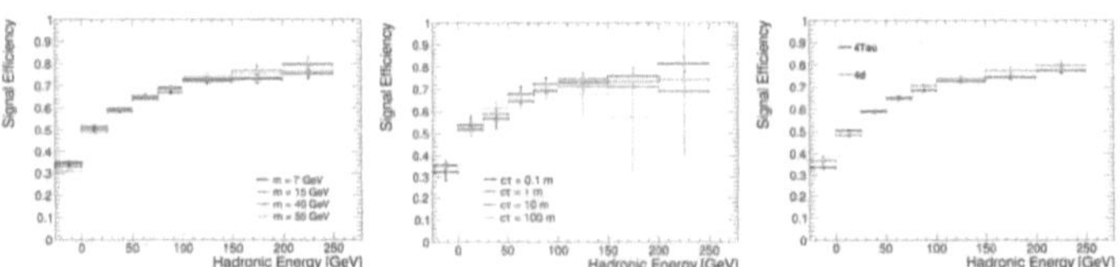

Figure 6.7: The efficiency of $N_{\text{stations}} > 1$ in region B, comparing different LLP masses (left), lifetimes (center), and decay modes (right). The different mass samples in left plot consists of the sum of all available $c\tau$ points (0.1–100 m). The different $c\tau$ samples in the center plot assumes an LLP mass of 15 GeV. The different decay mode samples in the right plot consists of all available mass (7–55 GeV) and $c\tau$ (0.1–100 m) points available. The first hadronic energy bin corresponds to LLPs that decayed leptonically with 0 hadronic energy. No dependence on LLP mass, lifetimes, and decay modes observed.

6.3 Simulation and Recast Strategy

The event generation and detector simulation framework, the recast procedure and its validation, followed by signal selections, signal and background yield estimate, and statistical analysis to evaluate the upper limits are detailed in this section.

6.3.1 Event Generation

Signal events are generated using MADGRAPH5 v2.9.3 [144], and parton shower and hadronization are performed with PYTHIA v8.244 [122], while keeping the LLP stable. Samples with different jet multiplicities are merged according to the MLM algorithm [145, 146]. Generator-level cuts are applied to the events to increase the sample size in the high p_T^{miss} phase space. Additional details on the samples, including the specific generator-level cuts, are given in Section 6.4 for each of the benchmark models considered in this work.

To efficiently decay the LLP, we used the fact that the reconstruction efficiency parameterization provided by the CMS search is spatially binned in two regions with simple shapes (defined by the intersection of ranges in the radial direction, r, longitudinal direction, z, and pseudorapidity, η), for which the probability of decaying inside a region can be computed analytically. For a region determined by the conditions

$$\eta_0 \le \eta \le \eta_1, \qquad r_0 \le r \le r_1, \qquad z_0 \le z \le z_1, \tag{6.1}$$

the probability, P, to decay inside the region for a particle traveling with momentum

p^μ and proper decay length $c\tau$ is

$$P = e^{-y_{min}} - e^{-y_{max}} \tag{6.2}$$

where

$$y_{min} = \frac{1}{\beta\gamma c\tau}\,max\Big(\,min\,(z_0\coth\eta, z_1\coth\eta), r_0\cosh\eta\Big), \tag{6.3}$$

$$y_{max} = \frac{1}{\beta\gamma c\tau}\,min\Big(\,max\,(z_0\coth\eta, z_1\coth\eta), r_1\cosh\eta, \beta\gamma c\,t_{cut}\Big(\beta\gamma + \sqrt{1+(\beta\gamma)^2}\Big)\Big), \tag{6.4}$$

with η being the pseudorapidity and $\beta\gamma = |\vec{p}|/m$. A timing requirement is also introduced such that $t_{\text{decay}} - d_{\text{decay}}/c < t_{\text{cut}}$ where t_{decay} and d_{decay} are the decay time and distance from the origin, respectively, and t_{cut} is the upper limit on the cluster time of 12.5 ns implemented in the CMS analysis.

The LLPs generated in the events are made to decay at fixed positions within each region using PYTHIA, as the presence of the decay vertex in a given region is the only geometrical information used by the detector simulation. For a given proper decay length $c\tau$, the probability for the LLP to decay in a given region is assigned to the event as an event weight. The decay probability for each LLP in an event is independent, so the event-level probability is the product of the decay probability of the LLPs in an event. Therefore, for each input event with an undecayed LLP, the decay program generates as many decayed events as the number of regions intersected by the LLP trajectory. The multi-weight capabilities of the HepMC event format is used to perform a scan in $c\tau$ without having to reprocess the events. In case there are multiple LLPs in the same event (as in the case of Higgs decays to pairs of dark vectors or scalars), LLP decays outside the signal regions are also included. In the case of LLP decays in the inner detector, where precise knowledge of the decay vertex position is used by the detector simulation, the decay vertex position is generated according to the decay probability distribution instead of a fixed number.

The events are subsequently passed to a simplified detector simulation based on DELPHES v3.4.2 [147] using the publicly available CMS configuration card for the reconstruction of prompt objects supplemented by a dedicated module, discussed in Section 6.3.2, simulating the LLP decay reconstruction and selection using the information provided by HEPData entry of the CMS search, as described in Section 6.2.

6.3.2 Detector Simulation with Dedicated Delphes Modules

The response of the CMS detector is simulated using Delphes [147]. The simulation uses the CMS detector configuration card [148], producing a set of standard particle flow (PF) candidates. The simulation of the clusters of hits in the CMS cathode strip chamber (CSC) of the endcap muon detector is performed using a dedicated Delphes module developed [149] using the parameterized detector response functions discussed in Section 6.2 and provided in the HEPData entry [143], associated with the CMS search result [1].

The simulation of cluster-level selection efficiencies are divided into three components. The first component is the cluster efficiency, which includes the cluster reconstruction efficiency, muon veto, active veto, time spread, and N_{hits} cut efficiency, as discussed in Section 6.2. Building upon the existing `Efficiency` module that is already used by all other PF candidates, I implemented a dedicated `CscClusterEfficiency` module in Delphes which encodes this parameterized function. The second component is cluster identification efficiency. A new `CscClusterID` module following the code function provided by the CMS HEPData entry is implemented.

The third component includes the model-dependent cluster time requirement, jet veto, and the $\Delta\phi(\vec{p}_T^{\text{miss}}, \text{cluster})$ requirement. The values of the three variables are calculated using generator-level information and saved as part of the `CscCluster` class. Specifically, the cluster time is determined by calculating the LLP travel time from the production to the decay vertex in the lab frame. The jet veto is implemented by requiring no PF jets with $p_T > 10$ GeV within a cone of $\Delta R = 0.4$ around the generator-level LLP direction. Finally, $\Delta\phi(\vec{p}_T^{\text{miss}}, \text{cluster})$ is computed as the azimuthal angle difference between the LLP direction and $\vec{p}_T^{\text{miss}}$, simulated using the standard Delphes modules. All these requirements are then made at a later stage of the analysis workflow.

Finally, the standard CMS configuration card from Delphes is modified to include the `CSCClusterEfficiency` and `CSCClusterID` module in the processing sequence. The modules require only generator-level LLP information and can be used for the recasting of the result for any model with LLPs. The implementation can be found in [149].

6.3.3 Analysis Strategy

In this section, we discuss the procedure used to recast the CMS Run 2 results and to make sensitivity projections for high-luminosity LHC (HL-LHC) [150] considering several different search strategy proposals.

For the recasting of the Run 2 results, the exact same selection and cuts are used as in the CMS paper. For all the standard PF objects, the default CMS configuration card from DELPHES is used, which has been validated [147] to reproduce the object resolutions from Run 2. The $p_{\mathrm{T}}^{\mathrm{miss}}$ calculation implemented in DELPHES is accurate when the LLP decays outside of the calorimeters that it's treated as invisible and when the LLP decays sufficiently close to the interaction point that the energy of the decay particles are measured by the calorimeters. In models where there is only one LLP in the event the LLP is required to decay in the muon detector, so the LLPs are treated consistently as invisible in both DELPHES and CMS simulations. For models with two or more LLPs per event, in the parameter space explored in this reinterpretation study, this approximation for the $p_{\mathrm{T}}^{\mathrm{miss}}$ calculation leads to a systematic error of 20% or less on the selection efficiency. For the CSC cluster objects, the `CSCClusterEfficiency` and `CscClusterID` modules are used to select clusters that would pass the corresponding selections. We then apply the $\Delta\phi(\vec{p}_{\mathrm{T}}^{\mathrm{miss}}, \text{cluster}) < 0.75$ selection, jet veto, and time cut for CSC clusters. The number of signal events passing these signal selections is used to calculate the signal yield. Finally, by using this estimate for the signal yield with the background yield (2 ± 1 background event) and observed data (3 observed events) obtained in the CMS analysis, the upper limits are calculated. The detailed limit calculation is discussed in Section. 6.3.4.

To further inform experimental studies and compare to other proposed LLP experiments, the sensitivity of this analysis is projected to the Phase 2 conditions, which is when the CMS detector will be upgraded to cope with HL-LHC. To simulate the effect on signal yield from the increased number of pileup interactions during Phase 2, the mean pileup number in the CMS configuration card is increased from 32 to 200. We assume that improved pileup mitigation algorithms and upgraded detectors, including the MIP Timing Layer (MTD) [151], will be able to mitigate the impact of the additional pileup on the object reconstruction and resolution. However, due to the larger number of jets from pileup interactions, the probability that a CSC cluster in the signal region is accidentally matched to and vetoed by a pileup jet with $p_T > 10$ GeV is 20% higher. Signal region CSC clusters are concentrated in the region with $|\eta| < 1.6$, while most pileup jets are concentrated in the high η region, so the increase in the probability of accidental matching is only 20%. Therefore, simple projection for Phase 2 constraints can be derived by scaling the signal (and background) yield by the increased integrated luminosity, and applying an 80% correction to the signal yield per cluster due to larger number of pileup jets while assuming the same efficiency and resolution for all PF candidates.

However, this simple recasting strategy significantly underestimates the potential sensitivity of a Phase 2 analysis. Realistically, given the larger dataset, one would apply tighter cuts to achieve near zero background. Therefore, a second recasting strategy is considered, where a tighter N_{hits} selection is applied until the expected background reaches zero. To estimate the signal and background yield with a tighter N_{hits} cut, the N_{hits} distributions in the auxiliary materials from the CMS analysis is used. The N_{hits} distribution for the background is fitted with an exponential function to extrapolate the background yield at higher N_{hits} cuts. A $N_{\mathrm{hits}} > 210$ requirement would suppress the background yield to below 1 for the expected Phase 2 integrated luminosity of 3 ab^{-1}. Increasing the cut from 130 to 210 would give a signal efficiency of about 80%. Therefore, for this recast strategy, the signal yield is scaled by an additional 0.8 with respect to the simple recasting strategy previously described and assumed a background yield of 0.2. The majority of the background comes from low p_T kaons and pions from the recoil, pileup, or underlying events that produce the clusters. The rate of low p_T particles from pileup would increase linearly with the number of pileup, but the background rate from the recoil in the main collision would not change with higher pileup. Therefore, we produce a conservative estimate of the effect of higher pileup by increasing the normalization of the N_{hits} distribution by the increase in number of pileup and recomputing the N_{hits} threshold. The estimated effect on the signal yield is at most 20%.

Finally, we consider two different search strategies that would be enabled by a new dedicated Level-1 and High Level Trigger targeting this signature starting from the beginning of Run 3. The trigger selects for events with at least one CSC cluster with a large number of hits and is already in operation and validated, as described in more detail in Chapter 7.

The first strategy with the new trigger targets models with at least two LLPs per events. In this strategy, we remove the high p_T^{miss} selection (which was necessary during Run 2 to trigger the events) and require at least two CSC clusters. In addition, we remove the requirement of at least one jet and $\Delta\phi(\vec{p}_T^{\mathrm{miss}}, \mathrm{cluster}) < 0.75$ that has high signal efficiency only in the high MET phase space. For this strategy, due to the double cluster requirement, we assumed that zero background can be achieved.

The second strategy with the new trigger is only implemented for the heavy neutral lepton model and we project the sensitivity to end of Run3 and end of Phase 2, corresponding to integrated luminosity of 300 fb^{-1} and 3 ab^{-1}, respectively. In this strategy, we no longer need to impose the high $p_{\mathrm{T}}^{\mathrm{miss}}$ threshold for triggering, so we reduce the $p_{\mathrm{T}}^{\mathrm{miss}}$ requirement from 200 GeV to 50 GeV, increasing the signal acceptance by three orders of magnitude. The rate of the main background, W+jets production, consequently increases by the same factor. We suppress the background to near negligible levels again by increasing the N_{hits} to 290 (370) resulting in a background yield of 0.2 for a dataset with an integrated luminosity of 300 fb^{-1} (3 ab^{-1}). We find that the signal yield increase due to the new trigger significantly offsets the 40% (50%) decrease in signal yield due to the $N_{\mathrm{hits}} > 290$ (370) with respect to the nominal N_{hits} threshold at 130.

For all the analyses discussed here, we assigned 20% signal systematic uncertainty which is of the same order of signal systematic for the CMS result. There it is dominated by missing higher order QCD corrections, which have a size of 21% for the gluon fusion production mode. The background uncertainty is dominated by statistical uncertainty, and we assign no additional background systematic uncertainty.

The result of our reinterpretation for Run 2 and the projection for Run 3 and Phase 2, are shown for all the benchmark models in Section 6.5.

6.3.4 Limit Calculation and Validation

Given the signal and background yield estimate, we evaluate the 95% confidence level (CL) limits using the "modified frequentist criterion" $\mathrm{CL_s}$ [141] for each point in the parameter space. The recasting procedure has been validated against the CMS exclusion limits, by recasting the twin Higgs model used in the CMS paper. The observed 95% CL upper limits on the branching fraction $\mathcal{B}(\mathrm{H} \to \mathrm{SS})$ for 15 GeV LLP as functions of $c\tau$ for the $S \to d\bar{d}$, $S \to b\bar{b}$, and $S \to \tau\bar{\tau}$ decay modes were compared against the CMS results, as shown in Figure 6.8. The limits evaluated using the fast simulation from DELPHES agree with the CMS result to within 30% for all lifetimes evaluated. Additionally, we also validate the $\Delta\phi(\vec{p}_{\mathrm{T}}^{\mathrm{miss}}, \mathrm{cluster})$ distributions for 15 GeV and 1 m proper decay length LLP decaying to $d\bar{d}$ against the auxiliary materials provided by CMS, as shown in Figure 6.9. The signal yield when requiring $\Delta\phi(\vec{p}_{\mathrm{T}}^{\,\mathrm{miss}},\mathrm{cluster}) < 1$ agrees within 7%.

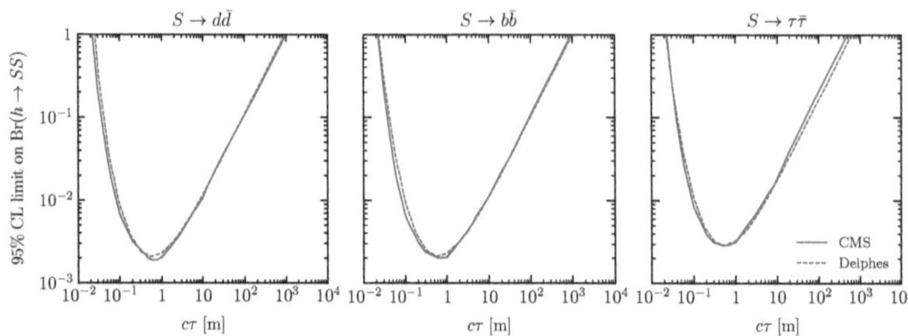

Figure 6.8: Comparison of the 95% CL upper limits on the branching fraction $\mathcal{B}(\mathrm{H} \to \mathrm{SS})$ as functions of $c\tau$ derived with the standalone workflow (dashed lines) and the CMS search (solid lines). In deriving these limits we have considered a 15 GeV LLP decaying into d-quark pairs (left), b-quark pairs (center), and τ pairs (right). The limits from this work are shown to agree with the CMS search to within 30%.

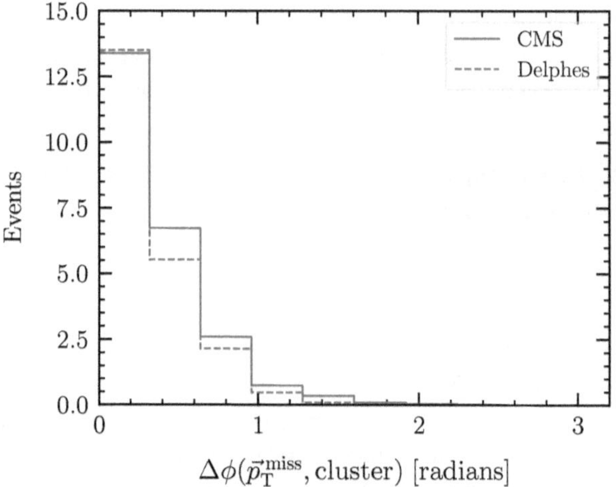

Figure 6.9: Comparison of the $\Delta\phi(\vec{p}_{\mathrm{T}}^{\,\mathrm{miss}},\mathrm{cluster})$ distributions derived with the standalone workflow (dashed line) and the CMS search (solid line) for a 15 GeV and 1 m proper decay length LLP decaying into d-quark pairs with $\mathcal{B}(\mathrm{H} \to \mathrm{SS}) = 0.01$. The signal yield when requiring $\Delta\phi(\vec{p}_{\mathrm{T}}^{\,\mathrm{miss}},\mathrm{cluster}) < 1$ agrees within 7%.

6.4 Benchmark Models

In this section, we briefly describe the benchmark models considered in this work. Each of these models has been chosen to showcase the strengths and limitations of the current analysis in concrete examples exhibiting different kinematics and signal topologies. Specifically, we want to investigate what happens to the analysis reach when lower values of LLP masses are chosen, when the LLP energy, E_{LLP}, is reduced, or when the LLP momentum is not correlated in magnitude or direction with the missing transverse energy. We also want to investigate what happens when there are multiple LLPs produced roughly in the same direction, potentially leading to failed isolation cuts in a non-trivial way.

Concretely, the models we consider are:

- Exotic Higgs decays into dark photons or light scalars. These are the closest models to the one considered in the original CMS analysis and are characterized by a production rate decoupled from the exclusive decay channels and LLP lifetime. Besides being commonly chosen benchmarks to compare the performance of different experiments in LLP searches, these benchmarks will allow us to probe the reach for LLP masses lighter than those presented in the CMS analysis, for a fixed production rate and using more realistic decay modes.

- Axion-like particles (ALPs) coupled to SM gauge bosons. In this model, the coupling to the SM is provided by a dimension five operator. A single parameter (the ALP decay constant) controls the production rate and lifetime. These models are characterized by a production cross section which is enhanced for energetic LLPs, irrespective of the light LLP mass.

- Inelastic Dark Matter (DM). In these models, the LLP is provided by an almost degenerate partner of the DM, and the amount of energy carried out by the LLP is controlled by the DM-LLP mass splitting and decoupled from the p_T^{miss}. This allows us to probe the reach in the low E_{LLP} region while allowing the other selection requirements to be passed without much of a penalty.

- Confining Hidden Valley models where jets of LLPs are produced in perturbative hidden showers, analogously to the case of QCD. This benchmark allows us to study the impact of the jet veto in models where multiple LLPs are produced in the same detector region. This is the same as the dark shower models described in Chapter 5. This reinterpretation study done on the first endcap-only search motivated the additional interpretation in the second combination paper.

- Heavy Neutral Leptons (HNL) that are produced in association with prompt leptons from a W boson. The production cross section and the HNL lifetime are both controlled by the mixing angle between the HNL and SM leptons.

6.4.1 Light Scalar Singlet

The most minimal extension of the SM is provided by adding a real scalar singlet (S) that mixes with the SM Higgs through renormalizable operators. The Lagrangian for this model reads [152]:

$$\mathcal{L}_{SH} = \mathcal{L}_{SM} + \mathcal{L}_{DS} - \left(A_{HS}\,\hat{S} + \lambda_{HS}\,\hat{S}^2 \right) \hat{H}^\dagger \hat{H} \tag{6.5}$$

where $\mathcal{L}_{SM}$ is the SM Lagrangian, H is the complex Higgs doublet, and the dark sector Lagrangian is given by

$$\mathcal{L}_{DS} = \frac{1}{2}\partial_\mu \hat{S}\,\partial^\mu \hat{S} - \frac{\mu_S^2}{2}\hat{S}^2 + \ldots \tag{6.6}$$

where we have omitted possible self-interactions of the scalar singlet, which we assume have been chosen in such a way that S does not have a vacuum expectation value. Here and in the following, we indicate with a hat the original fields with non-canonical kinetic terms, before any field redefinition is performed.

After electroweak symmetry breaking the Higgs scalar, $\hat{h}$, mixes with the singlet $\hat{S}$. The resulting physical states, h and S, obtained by diagonalizing the mass matrix, are given by the linear combination

$$\begin{pmatrix} h \\ S \end{pmatrix} = \begin{pmatrix} \cos\theta & \sin\theta \\ -\sin\theta & \cos\theta \end{pmatrix} \begin{pmatrix} \hat{h} \\ \hat{S} \end{pmatrix}, \tag{6.7}$$

where the mixing angle is controlled by the parameters A_{HS}, and explicitly given by

$$\tan\theta = \frac{x}{1 + \sqrt{1 + x^2}} \qquad x = \frac{2v A_{HS}}{\mu_H^2 - \mu_S^2 - \lambda_{HS}v^2}, \tag{6.8}$$

with v being the Higgs vacuum expectation value (vev), and $\mu_H^2 = \lambda_H v^2$ with λ_H the Higgs quartic coupling. The mass eigenvalues can also be expressed in terms of the small parameter x as

$$m_{h,S}^2 = \left(\frac{\mu_H^2 + \mu_S^2 + \lambda_{HS}v^2}{2} \right) \pm \left(\frac{\mu_H^2 - \mu_S^2 - \lambda_{HS}v^2}{2} \right) \sqrt{1 + x^2}, \tag{6.9}$$

which for $x \ll 1$ reduce to $m_h^2 \simeq \mu_H^2$ and $m_s^2 \simeq \mu_S^2 + \lambda_{HS} v^2$.

Due to the mixing in Equation 6.7, S inherits all couplings of the SM Higgs, modulo a suppression factor, $\sin\theta$, which is controlled by the parameter A_{HS}. Therefore, the decay width of the singlet can be obtained by rescaling that of a SM Higgs of the same mass. Specifically, we follow references [120, 142] to derive the singlet branching ratios used in this work.

In general, the production cross section is fixed by a combination of the parameters A_{HS} and λ_{HS}. The former controls the production via the $b \to s$ penguin diagram (allowed for $m_S < m_B - m_K$) [153–155] and $s \to d$ penguins for (allowed for $m_S < m_K - m_\pi$). The latter fixes the double S production via the $b \to s$ penguin diagram with an off-shell SM Higgs [156], or through direct Higgs decay. In the presence of a non-vanishing λ_{HS} (and for $2m_S < m_h$) the Higgs can decay into a couple of S with a width given by [152]

$$\Gamma_{h \to SS} = \frac{\lambda_{HS}^2 v^2}{8\pi\, m_h} \sqrt{1 - 4m_S^2/m_h^2}\,. \tag{6.10}$$

When $b \to s$ transitions dominate the production channel, decay and production are controlled by the same parameter, θ, and the model parameter space is given by $\{\sin\theta, m_S\}$. However, the analysis discussed in this work has no reach for the products of $b \to s$ transitions, as the LLPs would be mostly produced inside (or near) b-jets and fail isolation cuts, additionally the high p_T^{miss} requirement also suppresses the signal efficiency. Therefore, we will concentrate on the limit where the production is dominated by Higgs decays to two S, which is controlled by the parameter λ_{HS}. Therefore, production and decay channels will be decoupled and the model parameter space given by $\{\lambda_{HS}, \sin\theta, m_S\}$.

Concretely, we generated events for Higgs production from gluon fusion in association with up to two jets and decayed the Higgs to two scalars. No generator level cuts are imposed and the Higgs p_T distribution is reweighed to the next-to-next-to-leading order (NNLO) prediction.

We conclude by noticing that, given $m_S^2 \simeq \mu_S^2 + \lambda_{HS} v^2$, some level of fine-tuning is required for $m_S^2 < \lambda_{HS} v^2$. Measuring the degree of fine-tuning in terms of the parameter $\Delta \equiv m_S^2/(\lambda_{HS} v^2)$, we can write the branching ratio for the exotic decay $h \to SS$ as

$$\mathcal{B}(h \to SS) \simeq \frac{\Gamma_{h \to SS}}{\Gamma_h^{SM}} \simeq 6 \cdot 10^{-3} \left(\frac{m_S}{2.5\,\mathrm{GeV}}\right)^4 \left(\frac{0.1}{\Delta}\right)^2, \tag{6.11}$$

where Γ_h^{SM} is the total SM Higgs width.

6.4.2 Abelian Hidden Sector

The next benchmark model that we consider consists of extending the SM by adding a *dark* U(1) gauge symmetry which is spontaneously broken by a dark Higgs field, S. The dark U(1) is mediated by a *dark* photon, X, which kinetically mixes with the SM hypercharge:

$$\mathcal{L}_{SH} = \mathcal{L}_{SM} + \mathcal{L}_{DS} - \lambda_{HS}\,\hat{S}^2 \hat{H}^\dagger \hat{H} - \frac{\epsilon}{2\cos\theta_W}\hat{X}_{\mu\nu}\hat{B}^{\mu\nu}, \tag{6.12}$$

where $\hat{B}_{\mu\nu}$ and $\hat{X}_{\mu\nu}$ are the field strengths of the hypercharge and the new U(1) gauge group, respectively. The dark sector Lagrangian is written as:

$$\mathcal{L}_{DS} = -\frac{1}{4}\hat{X}_{\mu\nu}\hat{X}^{\mu\nu} + \mu_S^2\hat{S}^2 - \lambda_S\hat{S}^4 + |(\partial_\mu + ig_D\hat{X}_\mu)\hat{S}|^2. \tag{6.13}$$

As before, we indicate with a hat the original fields with non-canonical kinetic terms, before any field redefinition is performed. The dark U(1) is spontaneously broken by the vev of the dark Higgs, $\langle S \rangle = v_S/\sqrt{2}$, which generate a mass for the dark photon $m_{X,0} = g_D v_S$.

After electroweak symmetry breaking the kinetic mixing between the dark photon and the hypercharge induces a coupling of the dark photon to the SM fermions which, in the $m_X^2 \ll m_Z^2$ limits, reads

$$\mathcal{L}_{Xf\bar{f}} = \epsilon e\, Q_f X_\mu \bar{f}\gamma^\mu f, \tag{6.14}$$

where Q_f is the fermion electric charge. This coupling, controlled by the small parameter ϵ, provides the decay channel in visible states for the dark photon. Specifically, we compute the dark photon branching ratios by using the package provided in [157].

The diagonalization of the scalar sector proceeds similarly to what was discussed in the previous section, with the only difference that now we are interested in the regime where $m_S \gg m_h$, so that the dark Higgs decouples from the phenomenology of the model. Given the non-vanishing coupling between S and X, the mixing between the SM and dark Higgs generates a non-zero hXX coupling which gives rise to the exotic Higgs decay $h \to XX$, with a width given by

$$\Gamma(h \to XX) = \frac{\lambda_{HS}^2}{32\pi}\frac{m_h v^2}{m_S^2}\sqrt{1 - \frac{4m_X^2}{m_h^2}}\frac{(m_h^2 + 2m_X^2)^2 - 8(m_h^2 - m_X^2)m_X^2}{m_h^4}. \tag{6.15}$$

In the limit of small ϵ (which will be the relevant limit for our analysis), this dominates over Drell-Yan and $h \to ZX$ production and becomes the dominant dark photon production channel. In this limit the decay channel, controlled by ϵ, and the production channel, controlled by λ_{HS}, are decoupled. The model parameter space is given by $\{\epsilon, \lambda_{HS}, m_X\}$. The event generation for this benchmark was performed similarly to the light scalar singlet case.

6.4.3 Inelastic Dark Matter

Inelastic Dark Matter (iDM) models are characterized by a DM candidate that couples with the SM only through interactions with a nearly degenerate state. A simple realization of this scenario can be obtained by adding to the model discussed in the previous section a Dirac pair of Weyl fermions, η and ξ, that couple to the dark photon, X, with opposite charges. As before, the Higgs provides a source of U(1) breaking, generating a mass for the dark photon and a Majorana mass, δ, for the two Weyl fermions. A Dirac mass, m_D, involving the two Weyl fermions is also allowed, so that at energies below the dark U(1) breaking scale, the mass terms for the dark fermions are

$$\mathcal{L} \supset -m_D\, \eta\xi - \frac{\delta}{2}(\eta^2 + \xi^2) + \text{h.c.}. \tag{6.16}$$

For a technically natural small Majorana mass, these mass terms can be perturbatively diagonalized to give the physical states

$$\chi_1 \simeq \frac{i}{\sqrt{2}}(\eta - \xi) \qquad \chi_2 \simeq \frac{1}{\sqrt{2}}(\eta + \xi), \tag{6.17}$$

which have nearly degenerate masses $m_{1,2} \simeq m_D \pm \delta$. These mass eigenstates couple off-diagonally to the dark photon, i.e.,

$$\mathcal{L} \supset ie_D \hat{X}_\mu\, \bar{\chi}_1 \gamma^\mu \chi_2 + O\left(\frac{\delta}{m_D}\right), \tag{6.18}$$

where we have written $\chi_{1,2}$ as Majorana spinors using four-component notation. Therefore, if $m_X > m_1 + m_2$ and $\alpha_D \gg \epsilon\alpha_{em}$, once produced dark photons decay into $\chi_1\chi_2$ pairs with a rate given by $\Gamma_{X \to \chi_1\chi_2} \simeq \alpha_D m_X$, and provide the dominant production channel for $\chi_1\chi_2$ pairs at LHC. For the values of ϵ we are interested in this analysis, the dominant production channel for dark photons is provided by Drell-Yan processes and scales as ϵ^2.

The lightest state, χ_1, is stable and once produced leaves the detector as missing energy; χ_2 can decay into χ_1 plus a pair of SM particles through an off-shell dark photon, possibly leaving a detectable signature. The rate for decays with leptonic final states is given by [158]:

$$\Gamma_{\chi_2 \to \chi_1 l \bar{l}} = \epsilon^2 \alpha_{em} \alpha_D \int_{4m_l^2}^{(m_1 \Delta)^2} ds \frac{|\vec{p}_1|(m_1^2 \Delta^2 - s)(2s + m_1^2(2 + \Delta)^2)(s + 2m_l^2)(s - 4m_l^2)^{1/2}}{6\pi m_2^2 s^{3/2}(s - m_X^2)^2}$$

(6.19)

where s is the invariant mass of the lepton pair, $\vec{p}_1$ is the momentum of χ_1 in the rest frame of χ_2, and we have introduced the dimensionless parameter $\Delta \equiv (m_2 - m_1)/m_1$. The rate for decays involving hadronic final states can be derived by setting $m_l = m_\mu$ and multiplying the integrand of Equation 6.19 by the experimentally measured quantity $R(s) \equiv \sigma(e^+ e^- \to \text{hadrons})/\sigma(e^+ e^- \to \mu^+ \mu^-)$.

For this benchmark, events were generated using a MADGRAPH5 Z' model for X via production in association with up to three jets. A generator level cut $p_T > 100\,\text{GeV}$ was applied on the X, as the truth-level p_T^{miss} is given by p_T of the Z' (its decay products are one DM particle and an LLP decaying in the muon chambers).

6.4.4 Axion-like Particles

Axion-like particles (ALPs) extend the axion scenario to include any pseudoscalar particle that couples to the SM through dimension five operators. The naturally suppressed couplings make them a natural candidate for LLP searches. The general Lagrangian for these kinds of models is given by

$$\mathcal{L} = \mathcal{L}_{SM} + \frac{1}{2}\left(\partial_\mu a\right)^2 - \frac{1}{2}m_a^2 a^2 + \frac{c_q^{ij}}{2f}(\partial_\mu a)\bar{q}_i \gamma^\mu \gamma^5 q_j + \frac{c_\ell^{ij}}{2f}(\partial_\mu a)\bar{\ell}_i \gamma^\mu \gamma^5 \ell_j$$
$$+ \frac{a}{4\pi f}\left(\alpha_s c_{GG}\, G_{\mu\nu}^a\, \widetilde{G}^{a,\mu\nu} + \alpha_2 c_{WW}\, W_{\mu\nu}^a\, \widetilde{W}^{a,\mu\nu} + \alpha_1 c_{BB}\, B_{\mu\nu}\, \widetilde{B}^{\mu\nu}\right) + \cdots \quad (6.20)$$

where $\widetilde{G}_{\mu\nu} = 1/2\,\epsilon_{\mu\nu\rho\sigma} G^{\rho\sigma}$ where $G^{\rho\sigma}$ is the gluon field strength, and similarly for $\widetilde{W}$ and $\widetilde{B}$. In the broken, phase the couplings to W and B bosons induce couplings to photons and Z-bosons which are given by:

$$c_{ZZ} = c_{WW} + c_{BB} \qquad c_{\gamma Z} = c_w^2 c_{WW} - s_w^2 c_{BB} \qquad c_{\gamma\gamma} = c_w^4 c_{WW} + s_w^4 c_{BB}. \quad (6.21)$$

In this work, we will focus on benchmark models in which the ALP couples only to gauge bosons ($c_q^{ij} = c_\ell^{ij} = 0$). Since the focus is on the production of energetic, isolated LLPs, this choice is sufficient to capture most of the dominant production modes at the LHC. Specifically, we will consider the three following scenarios: ALP coupled to W ($c_{WW} \neq 0$, $c_{GG} = c_{BB} = 0$), photophilic ALP ($c_{\gamma\gamma} \neq 0$, $c_{\gamma Z} =$

$c_{GG} = 0$), and ALP coupled to gluons ($c_{GG} \neq 0$, $c_{BB} = c_{WW} = 0$). The latter is a well-studied benchmark model in the context of light LLP searches, yielding the highest production rate at the LHC. The photophilic model chosen here is one of the (infinite) possible choices of UV-completion at LHC energies of the well-studied low-energy benchmark of "ALP coupled to photons". The conservative choice $c_{\gamma Z} = 0$ is to focus on the parameter region where the existing LEP bounds are the weakest. Finally, the $c_W \neq 0$ benchmark provides a better UV-motivated benchmark than the photophilic choice, where associated ALP production with all the gauge bosons is allowed.

For the ALP coupled to gluons, we generated events where the LLP is produced in association with up to 3 jets, and imposed a $p_T > 100$ GeV and a $|\eta| < 3$ generator-level cuts on the transverse momentum and pseudorapidity of the ALP. The MADGRAPH5 model used here has been described in [159], and we have only adapted the normalization of the couplings to the one used above. We did not include ALP production in the shower (i.e., where the ALP is produced at intermediate scales between the hard process collision and the QCD confinement scale) which was first estimated in [160], as there are not yet reliable event generators that can be used to keep track of the angular separation between the ALP and QCD jets (necessary for the jet veto requirements of the analysis). Therefore, for this benchmark, our limits should be considered conservative estimates for the reach of this CMS analysis, as they miss an important production channel. Production from meson mixing and meson decay was also neglected because it yields softer and non-isolated ALPs, for which this analysis has no sensitivity. For the lifetime and exclusive decay branching ratios of this benchmark, we used the estimates of [161].

For the case of the other two ALP benchmarks, we considered ALP production in association with either a W, a Z, or a photon and up to 2 extra jets. We kept the same generator-level pseudorapidity cut but lowered the p_T cut to 50 GeV as some of the p_T^{miss} can be produced by the decay products of the W and Z bosons. In these two benchmarks, the ALP decays predominantly into two photons.

6.4.5 Hidden Valley

Confining Hidden Valleys (HV) [98], with a perturbative evolution below the scale mediating the interactions producing hidden sector particles, are a generic hidden sector extension of the SM on which we can have some theoretical control based on our knowledge of QCD-like theories. In general, one expects jets of hidden sector partons to hadronize in HV particles, some of which may decay back into SM final states, potentially as LLPs. Still, large freedom exists in defining a specific model. From the field content of the hidden sector and its symmetries, to the portal interactions mediating both the production of HV states and their decay back to the SM [114]. Many studies of search strategies at the LHC have been performed for different incarnations of this paradigm [162].

In the context of this reinterpretation study, we choose one particular realization as an example model generating the LLP-jet signature, aiming at maximizing the multiplicity of LLP produced in a jet, while keeping a high level of simplicity of reinterpretation. Therefore the example chosen is by no means generic *per se*, although the lessons learned about the CMS analysis are. Specifically, we used the hidden-valley module [124] implemented in PYTHIA to generate events and choose a perturbative hidden sector with an $SU(N_c)$ asymptotically free gauge group with N_f hidden quark flavors, fixing $N_c = 3$ and $N_f = 1$. The choice of $N_f = 1$ is to guarantee the absence of stable hidden mesons, therefore reducing the amount of collider stable particles produced and maximizing the number of LLPs in a hidden jet. This has a drawback, namely the lack of knowledge of the mass spectrum of such a theory as it lacks chiral symmetry breaking which is an important handle used in lattice simulation. In particular, the mass ratio between the first (pseudo-)scalar, η_V, and vector, ω_V, resonances are poorly known, but expected to be $O(1)$ [163–165]. Again, motivated by maximizing LLP multiplicity, we choose $m_{\omega_V} = 2.5 m_{\eta_V} = \Lambda_{HV}$ and assumed that the lowest scalar state (which PYTHIA will not use in the hadronization of the HV partons) is also able to decay to pairs of η_V. In this way, vector resonances can promptly decay to pseudoscalar mesons η_V, which will be the LLPs. For portals, we decide to decouple production and HV meson decays so that we can study the effects of varying the LLP lifetime on the limits for a fixed production rate. Specifically, we will produce hidden quarks in Higgs decays and will decay back the hidden spin-0 mesons 100% into pair of photons. The latter choice is purely driven by the fact that the CMS analysis is not too sensitive to the relative amount of hadronic vs electromagnetic energy in LLP decays. At the same time, existing limits on light LLPs decaying to pair of photons are quite weak, so we can focus on reinterpreting this analysis without worrying about recasting other existing searches. From a model building point of view, these portals can be easily generated by introducing a heavy scalar and pseudoscalar states S and A, having Yukawa interactions with the HV vector-like quark q_V. The scalar S can then interact with the SM Higgs via a $|H|^2 S$ cubic interaction, generating a

Yukawa coupling between q_V and the SM Higgs and a q_V mass after electroweak symmetry breaking. At the same time, the pseudoscalar A can have a coupling to the SM photons $AF\tilde{F}$ which in turn will induce a small decay width for η_V via $\eta_V - A$ mixing.

6.4.6 Heavy Neutral Leptons

We begin by defining the *minimal HNL model* and then discuss its relation to two basic seesaw models: The classical seesaw type-I [166–170] and the inverse seesaw [171].

An HNL is defined by its charged and neutral current interactions with standard model leptons:

$$\mathcal{L}_{int} = \frac{g}{\sqrt{2}} V_{\alpha N_j}\, \bar{l}_\alpha \gamma^\mu P_L N_j W^-_{L\mu} + \frac{g}{2\cos\theta_W} \sum_{\alpha,i,j} V^L_{\alpha i} V^*_{\alpha N_j} \overline{N}_j \gamma^\mu P_L \nu_i Z_\mu + \text{h.c.} \quad (6.22)$$

Here, $V_{\alpha N_j}$ are free parameters, parametrizing the mixing angle of N_j and, in principle, one can add $j = 1, \cdots, n$ HNLs to the SM. In searches, one typically assumes there is only one HNL with a mass in the kinematically accessible region. Note, $V^L_{\alpha i}$ is the mixing among light neutrinos.

To $\mathcal{L}_{int}$ one has to add a mass term. This mass term could be either of Dirac or Majorana type. For Dirac HNLs, only lepton number conserving decays (LNC) are possible, whereas Majorana HNLs can have both LNC and lepton number violating (LNV) decays. Thus, for the same values of m_N and $V_{\alpha N_j}$, the decay width of a Majorana neutrino is twice that of a Dirac neutrino. For definiteness, in the numerical part of this work we use Majorana HNLs.

The study of HNLs is usually motivated by the observed neutrino masses, see for example [172] for a recent overview on the status of neutrino data. The minimal HNL model, on the other hand, takes the $V_{\alpha N_j}$ for $\alpha = e, \mu, \tau$ as free parameters and does not explain light neutrino masses. To make contact with neutrino data one needs to connect the HNL with some theoretical neutrino mass model.

The simplest possible model is the type-I seesaw. In seesaw type-I one adds three right-handed neutrinos to the standard model field content. The model generates the mass matrix for the six neutral states:

$$\mathcal{M}_{type-I} = \begin{pmatrix} 0 & m_D^T \\ m_D & M_R \end{pmatrix}. \quad (6.23)$$

Here m_D is Dirac mass matrix, while M_R is the Majorana mass matrix for the right-handed singlets. In seesaw type-I one can always choose to work in the basis where the latter is diagonal, $\hat{M}_R$. After diagonalization of Equation (6.23) the light neutrino masses and the mixing between the light (and mostly active) and heavy (mostly sterile) neutrinos is given by

$$m_\nu \;=\; -m_D^T \cdot M_R^{-1} \cdot m_D \cdots \tag{6.24}$$

$$V_{H-L} \;=\; m_D^T \cdot M_R^{-1} \cdots, \tag{6.25}$$

where the dots represent higher order terms. Note that the matrix elements of V_{H-L} correspond to the mixing angle parameters $V_{\alpha N_j}$, in Equation (6.22), but we use a different symbol to distinguish it from the "model-independent" HNL setup. For the seesaw type-I, one can find a simple reparameterization of the Dirac mass matrix [173]:

$$m_D = i\sqrt{\hat{M}_R} \cdot \mathcal{R} \cdot \sqrt{\hat{m}_\nu} \cdot U_\nu^\dagger. \tag{6.26}$$

Equation (6.26) guarantees that the seesaw parameters chosen always fit the input neutrino data. Here, U_ν is the mixing matrix observed in oscillation experiments, $\hat{m}_\nu$ and $\hat{M}_R$ are the eigenvalues for the light and heavy neutrinos, respectively, and $\mathcal{R}$ is an orthogonal matrix of three complex angles. The entries in $\mathcal{R}_i$ can be written in terms of $s_i \equiv \sin(z_i)$, with $z_i = \kappa_i \times e^{2i\pi\xi_i}$ [174]. It is straightforward to show that for all $s_i = 0$, the matrix V_{H-L} is given by:

$$(V_{H-L})_{ij} = (U_\nu^*)_{ij}\sqrt{\frac{\hat{m}_{\nu,i}}{M_{R,i}}}. \tag{6.27}$$

Thus, one expects that the mixing V_{H-L} is suppressed by light neutrino masses in type-I seesaw. Also, in this limit the branching ratios of the heavy singlets to the different SM generations are related to measured neutrino angles. However, for complex $\mathcal{R}$ one can find larger values of V_{H-L}, if one allows the different contributions in Equation (6.24) to nearly cancel against each other. Note that in this fine-tuned part of parameter space, Equation (6.26) is no longer valid, since 1-loop corrections to the seesaw become more important than the tree-level itself, see the discussion in [175]. While in this "cancellation region" one can have mixings large enough to be experimentally testable, it is not possible to find right-handed neutrinos decaying to only one SM lepton generation in this particular part of parameter space of the seesaw [176]. This conclusion, however, is valid strictly only for type-I seesaw.

Very different expectations for V_{H-L}^2 and N_i branching ratios are obtained for the inverse seesaw mechanism. (We concentrate on inverse seesaw here, but a similar discussion could be presented for the *linear* seesaw [177, 178].) In inverse seesaw [171], three additional singlets, denoted S, are added and the $(9,9)$ mass matrix is given by:

$$\mathcal{M}_{\text{ISS}} = \begin{pmatrix} 0 & m_D^T & 0 \\ m_D & 0 & M_R \\ 0 & M_R^T & \mu \end{pmatrix}. \tag{6.28}$$

Note that, in the limit $\mu \equiv 0$ the three active neutrinos are massless, i.e., lepton number is conserved. Thus, a small value of μ is technically natural. In this limit the 6 heavy states form three Dirac pairs with masses M_{R_i}. For $\mu \ll m_D \ll M_R$, the mass matrix for the lightest three states, the masses of the heavy states and the mixing to the heavy neutrinos are given as:

$$(m_\nu)_{\text{ISS}} \quad = \quad m_D^T \cdot M_R^{T-1} \cdot \mu \cdot M_R^{-1} \cdot m_D + \cdots \tag{6.29}$$

$$M_\pm \quad \simeq \quad \left(M_R + \left\{ m_D.m_D^T, M_R^{-1} \right\} \right) \pm \frac{1}{2}\mu \tag{6.30}$$

$$V_{H-L} \quad = \quad \frac{1}{\sqrt{2}} m_D^T \cdot M_R^{-1} + \cdots \simeq \sqrt{\frac{\hat{m}_\nu}{\mu}} \tag{6.31}$$

Here $\{a, b\}$ is the anti-commutator of a and b. The heavy states thus form "pseudo-Dirac" pairs, splitted by the small parameter μ. In the limit $\Gamma \ll \mu$, where Γ is the total decay width of the heavy state, the singlets behave as Majorana particles, while for the opposite limit $\mu \ll \Gamma$, the decays are all Dirac-like [174]. Heavy-light mixing in inverse seesaw is given by the same ratio of m_D and M_R as for seesaw type-I, but the relation of V_{H-L} to light neutrino masses is changed, thus the second relation in Equation (6.31) above. Clearly, the naive expectation is that for an inverse seesaw model, the mixing V_{H-L} is *enhanced by a factor* $\hat{M}_R \cdot \mu^{-1}$ relative to the seesaw type-I.

One can formulate a parameterization of m_D in terms of neutrino oscillation parameters, M_R, μ and $\mathcal{R}$ [175] in the same spirit as the Casas-Ibarra parameterization for the type-I seesaw [173]. For $\mathcal{R} = \mathbb{1}$ one obtains the second equation in Equation (6.31) above. The larger number of free parameters in the inverse seesaw, however, allows not only to easily find parameter space with much larger V_{H-L} than for the seesaw type-I, it also offers the possibility to break the relation $V_{H-L} \propto U_\nu^*$, shown in Equation (6.27). The simplest possibility to do so, is to choose both m_D and M_R diagonal. In this case, according to Equation 6.31 V_{H-L} will be diagonal, and therefore each of the three (pairs of) heavy singlets will decay to only one generation of SM leptons. Even in this case, the neutrino data can be correctly fitted easily as can be seen in the following expression derived from Equation 6.29:

$$\mu = M_R^T \cdot (m_D^T)^{-1} \cdot U_\nu^* \cdot \hat{m}_\nu \cdot U_\nu^\dagger m_D^{-1} \cdot M_R \tag{6.32}$$

In subsequent sections we will denote these theoretical scenarios as electron-type, muon-type, and tau-type HNL.

The above discussion, while by far not covering all theoretical possibilities, serves to show that from the point of view of neutrino model building, larger HNL mixing is expected in the inverse seesaw model. Moreover, discovering an HNL with "large" mixing, but coupling to only one generation of SM charged leptons would be a strong hint that the underlying neutrino mass model is not the simplest type-I seesaw. In the numerical part of this work, we will, however, use the minimal HNL model, treating $V_{\alpha N}$ simply as free parameters.

We use the FeynRules implementation for HNLs of Ref. [179] to generate events in Madgraph5 for $pp \rightarrow W$, with up to two jets and $W^\pm \rightarrow Nl^\pm$. We apply generator-level cuts on HNL kinematics, $p_T \geq 100$ GeV and $0.5 < |\eta| < 3$, in MadGraph5 to increase the statistics in the phase space regions selected by the CMS analysis. The leading order (LO) W boson production cross section and the shape of the W p_T spectrum are corrected to the best known theoretical prediction at NNLO [180].

6.5 Results

In this section, we present the results for the benchmark models discussed in the previous section. We present both the current constraints, derived from data collected from 2016 to 2018, corresponding to an integrated luminosity of 137 fb^{-1} and the projected constraints for Run 3 (for HNL only) and Phase 2.

For the HNL model, two different projections are considered for Run 3 and Phase 2 discussed in Section 6.3.3. Strategy 1 corresponds to the search with a higher N_{hits} cut and zero background and strategy 2 corresponds to the search with a dedicated trigger, lowered p_T^{miss} requirement of 50 GeV, and increased N_{hits} cut. The experimental sensitivity for the HNL minimal scenario in the $|V_{\alpha N}|^2$ vs m_N plane for $\alpha = e, \tau$ are shown in Figure 6.10.

The reinterpretation of muon-type HNLs is not considered in this work. Insufficient information is provided for the muon detector response for displaced muons produced in the muon detector to estimate the efficiency of the muon vetos for muon-

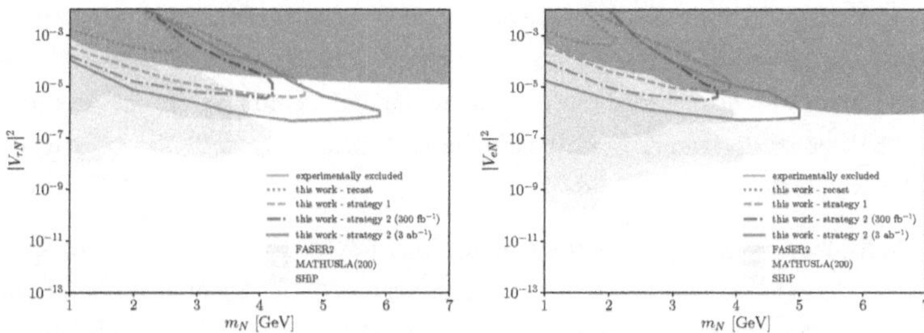

Figure 6.10: Recast and projected sensitivity of the different proposed search strategies with a displaced shower signature in the CMS muon system. The minimal HNL scenario is considered with mixings in the τ (left) and electron (right) sectors. The blue "recast" contour corresponds to a straightforward recast with the Run-2 dataset. Dashed green "strategy 1" is the same as the dot-dashed lines in the other interpretations, corresponding to an increased N_{hits} requirement. Strategy 2 corresponds to the strategy with the new trigger and lowered $p_{\text{T}}^{\text{miss}}$ requirement. Sensitivity of strategy 2 are shown for datasets with luminosity of 300 fb^{-1} and 3 ab^{-1} in black and brown, respectively. We compare our results with current constraints (gray shaded region) that come from limits from the DELPHI [181] and ATLAS experiments [182]. We also compare with projections from the proposed SHiP (yellow) [183], MATHUSLA (pink) [184], and FASER2 (blue) [185] experiments.

type HNLs that produces a displaced muon. On the other hand, only a few percent of signal events in electron and τ sector contain displaced muons passing the muon veto threshold of $p_{\text{T}} > 20\,\text{GeV}$ in the final state, so the impact on the sensitivity is negligible and is propagated as a source of signal systematic uncertainty.

As shown in Figure 6.10, the limits can reach mixing parameters several orders of magnitude smaller than current experimental bounds and are complementary to the proposed far detector experiments. The sensitivities in $|V_{\tau N}|^2$ can reach values down to $|V_{\tau N}|^2 \sim 5 \times 10^{-6}$ for $m_N \sim 5\,\text{GeV}$ using strategy 1 and 5×10^{-7} using strategy 2 with 3 ab^{-1}. For electron-type HNL, mixing parameter down to $|V_{eN}|^2 \sim 10^{-5}$ can be excluded for $m_N \sim 4\,\text{GeV}$ using strategy 1 for an integrated luminosity of 3 ab^{-1}. For the strategy 2, the limits can be improved up to $|V_{eN}|^2 \sim 10^{-6}$ for $m_N \sim 5\,\text{GeV}$ for the same integrated luminosity.

Finally, it is important to mention that only HNLs produced from W boson was considered. However, for masses $m_N \lesssim 5\,\text{GeV}$, the HNLs can also be produced from meson decays or τ lepton decays if it's kinematically allowed. These contributions to the HNL production are expected to be important for the limits obtained using strategy 2, which has a new dedicated displaced trigger, and does not require triggering on high p_T^{miss} nor a high p_T prompt charged lepton. The analysis of the sensitivities of a displaced shower signature on HNLs coming from meson decays will be studied in a future work, where the HNL mass range will also be extended to below 1 GeV.

For all other models, the different projections for Phase 2 are derived by using three different search strategies discussed in Section 6.3.3. Specifically, solid lines correspond to the search with the same selections as the CMS paper and a background rescaled according to the higher luminosity, dot-dashed lines correspond to the search with a higher N_{hits} cut and zero background (same as strategy 1 for HNL), and the dashed lines to the search with a dedicated trigger (without the high p_T^{miss} and isolation cuts, but requires two CSC clusters) and zero background.

The reach for the light scalar model with $\lambda = 1.6 \times 10^{-3}$ is shown in Figure 6.11. This choice of λ corresponds to an exotic Higgs branching fraction of $\mathcal{B}(\text{H} \to \text{SS}) = 0.01$, which is roughly the future reach for the Higgs branching into invisible final states. The present constraints are shown in the left panel, where for low masses below a few GeV the analysis probes a previously unconstrained region of the parameter space, while at higher masses the constraints are similar to that of the ATLAS search for displaced vertices in the muon chambers (indicated as ATLAS mu-ROI in Figure 6.11). In the right panel of Figure 6.11, the projections for Phase 2 and comparison with the projected constraints from other future experiments are shown. Due to the smaller distance to the interaction point (IP), the projected results of using MDS are complementary to and sensitivity to larger mixing angles compared to the other dedicated LLPs experiments that are positioned further away from the IP.

To give an idea of how the constraints depend on the value of $\mathcal{B}(\text{H} \to \text{SS})$, constraints for different values of $\mathcal{B}(\text{H} \to \text{SS})$ are shown in Figure 6.12. As seen in the figure, the current search can reach as low as $\mathcal{B}(\text{H} \to \text{SS}) \lesssim 3 \times 10^{-3}$, while the future Phase 2 search can reach $\mathcal{B}(\text{H} \to \text{SS}) \lesssim 3 \times 10^{-4}$. Alternatively, in Figure 6.13 we show the limits for a different slicing of the parameter space of this model, where the tree-level mass for S is absent and the LLP mass is fully controlled by λ_{HS} and therefore by $\mathcal{B}(\text{H} \to \text{SS})$. In this case, there is no tuning even for lower masses, but the production rate varies with m_S and searches for $H \to$ inv. set an upper bound on m_S. Finally, to compare with the results of the CMS analysis, in Figure 6.14 we report the present and future limits on $\mathcal{B}(\text{H} \to \text{SS})$ as a function of the scalar

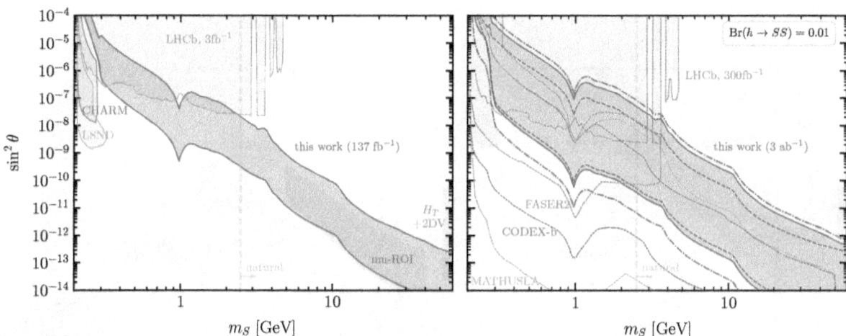

Figure 6.11: Constraints on light scalars produced in Higgs decays for $\mathcal{B}(\text{H} \to \text{SS}) = 0.01$. **Left:** Comparison of our current reach (red region) with existing limits from LHCb (orange) [186], LSND (azure) [187], reinterpretation [120] of the CHARM experiment (blue) [188], CMS HT +2DV search (green) [142, 189], and reinterpretation of ATLAS mu-ROI (purple) [142, 190]. **Right:** Projections of our constraints for a luminosity of 3 ab^{-1} (red region). The three red contours (solid, dashed, and dot-dashed) correspond to the three search strategies discussed in the main text (rescaled CMS analysis, dedicated trigger, and higher N_{hits}). We compare our results with current constraints (gray shaded region) and projections for MATHUSLA [191], CODEX-b [160], FASER2 [192], and LHCb 300fb^{-1} [193]. The constraints for the projections are shown between the dashed lines with the corresponding colors. The vertical orange line indicates the scalar mass below which the model needs to be fine-tuned (see discussion around Equation 6.11).

lifetime.

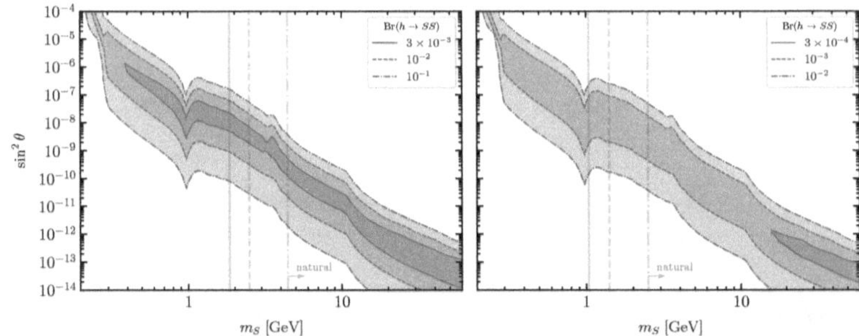

Figure 6.12: Our limits for the light scalar model for different values of $\mathcal{B}(\mathrm{H} \to \mathrm{SS})$. In the left panel we show the current reach, while in the right panel we present the $3\,\mathrm{ab}^{-1}$ projections assuming that the same selections of the original CMS analysis are used. As in the previous plot, the vertical lines indicate the scalar mass below which which tuning of more than 10% is present (see discussion around Equation 6.11).

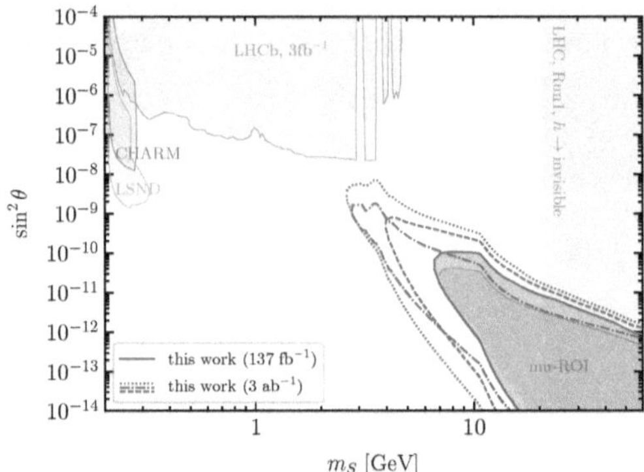

Figure 6.13: Constraints on the singlet scalar model in absence of a tree-level mass for S ($\mu_S = 0$). The solid red line shows the current constraints, while the other contours (dashed, dot-dashed, and dotted) show $3\,\mathrm{ab}^{-1}$ projections derived by using the three recast strategies discussed in Section 6.3.3 (rescaled CMS analysis, dedicated trigger, and higher N_{hits}). The other existing constraints appearing on the plot are the same of Figure 6.11.

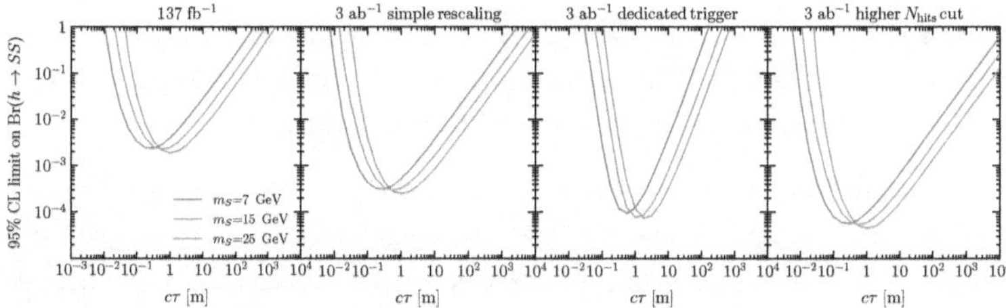

Figure 6.14: Upper limits on the branching fraction $\mathcal{B}(\mathrm{H} \to \mathrm{SS})$ as functions of $c\tau$. In the left panel, we report the current constraints set by our analysis. In the other panels, we show the projected constraints for a luminosity of $3\,\mathrm{ab}^{-1}$ derived from the three different search strategies discussed.

The constraints for the Abelian hidden sector are shown in Figure 6.15. As before, the value of the exotic Higgs branching ratio is fixed to $\mathcal{B}(h \rightarrow A'A') = 0.01$. It can be seen that our current constraints cover a mostly unconstrained region of the parameter space, except for the overlap with the ATLAS mu-ROI search at high masses. As for the scalar model, our projected constraints well complement dedicated LLPs searches due to the different distances from the IP. To investigate the lowest value of $\mathcal{B}(h \rightarrow A'A') = 0.01$ that can be probed, the present and future constraints for different values of the exotic Higgs branching are shown in Figure 6.16. As seen in the figure, the current search can reach as low as $\mathcal{B}(h \rightarrow A'A') = 3 \times 10^{-3}$, while the future Phase 2 search can reach $\mathcal{B}(h \rightarrow A'A') = 3 \times 10^{-4}$. This is consistent with what was found for the singlet scalar model and shows the independence of the analysis to the specific exclusive decay modes. The only significant differences are around resonance mixing with hadronic resonances, which differ between the scalar and vector LLPs, and affect the LLP lifetime and in the region between $200\,\text{MeV} \lesssim m \lesssim 300\,\text{MeV}$ where the 2μ final state, to which this analysis is not sensitive to, contributes to $\mathcal{O}(50\%)$ of the dark photon branching ratios.

The constraints for the three ALP models are shown in Figure 6.17 - 6.19. The mass dependence of the limit from the ATLAS mono-jet search for gluon-coupled ALP in Figure 6.17 was derived in this work as described in Appendix 6.A. For both the gluon (Figure 6.17) and electroweak (Figure 6.18 and Figure 6.19) coupled scenarios, we find that the reinterpretation of the CMS analysis covers new territory beyond previous monojet [207] and fixed target [188] searches while being complementary to dedicated LLP experiments. Moreover, one can expect the projections shown here to be underestimated, as dedicated searches using the fact the ALP is produced in association with a photon or a vector boson may allow us to relax some of the cuts and access softer LLPs that are produced with higher rates, extending the estimated limits towards higher ALP masses and decay constants.

We now turn to the inelastic DM model results. The reinterpretation of this model is fairly sensitive to the LLP energy, E_{LLP}, due to the small mass splitting, Δ. However, the cluster efficiency tables are not granular enough at low deposited energies (E_{em}, E_{had}) to resolve the turn-on shoulder of the 2D efficiency surface (the first bin is between 0 and 25 GeV). Therefore, our ability to reliably recast this model is hampered by the lack of knowledge about the minimal energy threshold for which the LLP visible decay products can produce $\mathcal{O}(20-30)$ charged particles emerging from a steel layer into the muon stations. To estimate this energy threshold, we impose

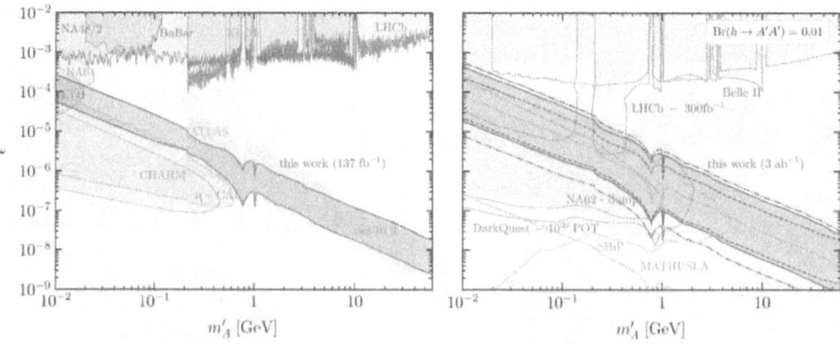

Figure 6.15: Constraints on dark-photons produced in Higgs decays for $\mathcal{B}(\mathrm{H} \to SS) = 0.01$. **Left:** Comparison of our current reach (red region) with existing limits from BaBar (blue) [194], KLOE (azure) [195], LHCb (purple) [196], NA48 (brown) [197], reinterpretation of ATLAS μ-ROI (yellow) [190], ATLAS search for displace dark-photon jets (yellow) [198], and beam dump experiments (orange, gray, green, pink) [188, 199–201]. Most of the experimental constraints appearing in this plot have been digitized with the help of `darkcast` [157]. **R**ight: Projections of our constraints for a luminosity of $3\,\mathrm{ab}^{-1}$ (red region). The three red contours (solid, dashed, and dot-dashed) correspond to the three search strategies discussed in the main text (rescaled CMS analysis, dedicated trigger, and higher N_hits). We compare our results with current constraints (gray shaded region) and projections for MATHUSLA (orange) [191], SHiP (azure) [183], DarkQuest (purple) [202, 203], NA62 in dump mode (green) [202, 204], LHCb upgrade (brown) [202, 205], and Belle II (blue) [206].

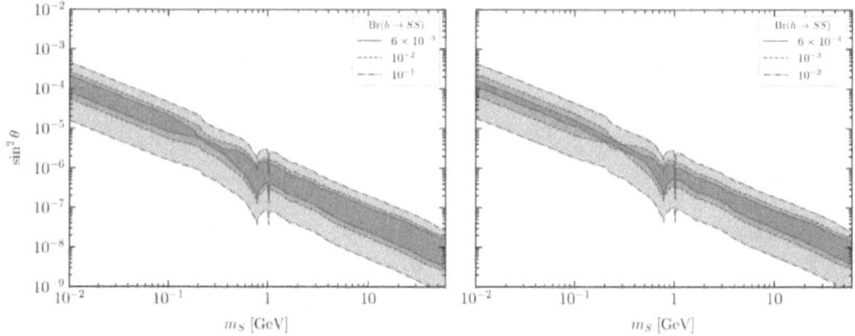

Figure 6.16: Our limits for the dark photon model for different values of $\mathcal{B}(\mathrm{H} \to SS)$. In the left panel we show the current reach, while in the right panel we present the projections for $3\,\mathrm{ab}^{-1}$ assuming the same selections of the original CMS analysis are used.

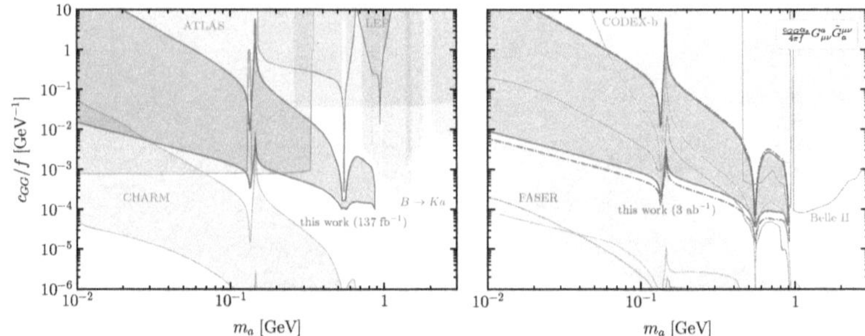

Figure 6.17:   Constraints on ALPs coupled to gluons. **Left:** Comparison of our current reach (red region) with existing limits from CHARM (orange) [160, 188], our reinterpretation of ATLAS (green) [207], LEP [161], and flavor probes (purple) [30, 208–211].   **Right:** Projections of our constraints for a luminosity of 3 ab^{-1} (red region). The solid and dot-dashed red contours correspond to the projections derived by using the same selections of the original CMS analysis, and the one derived by using a higher N_{hits} cut and assuming zero background. We compare our results with current constraints (gray shaded region) and projections for FASER (purple) [212], CODEX-b (orange) [160], and Belle II (green) [208, 209]

an additional cut $E_{LLP} > 5\,\text{GeV}$ (which is approximately the energy needed for an electron to produce $\mathcal{O}(20)$ charged particles at the shower development maximum). The constraints for this choice of cut and using the model parameters $\Delta = 0.005$, $\alpha_D = 0.1$, and $m_{A'} = 3m_1$, are reported in Figure 6.20. We see that the analysis covers previously unconstrained regions of the parameter space near the Z-resonance at $m_{A'} = 3m_1 = m_Z$. To further estimate the sensitivity of these results to the lower cut on the LLP energy, we show in Figure 6.21 the effect of varying it between 0 and 10 GeV.

In Figure 6.22 we report the limits on the exotic Higgs branching ratio $\mathcal{B}(h \rightarrow Q\bar{Q})$ for the hidden valley model with HV confining scale $\Lambda_{HV} = 20\,\text{GeV}$, which correspond to a pseudoscalar mass of $m_{\eta_V} = 8\,\text{GeV}$. Since in this model LLPs are produced within dark-showers in LLP jets, we expect the jet veto to reduce the sensitivity of the analysis. To quantify this effect, in the lower panels of Figure 6.22 we show the ratio of the signal efficiency of the CMS analysis divided by the signal efficiency of the same analysis without the jet veto. As expected, this ratio rapidly approaches zero for small LLP lifetimes, when it is more likely for multiple LLPs to decay within the inner detector regions and the calorimeters in front of the cluster

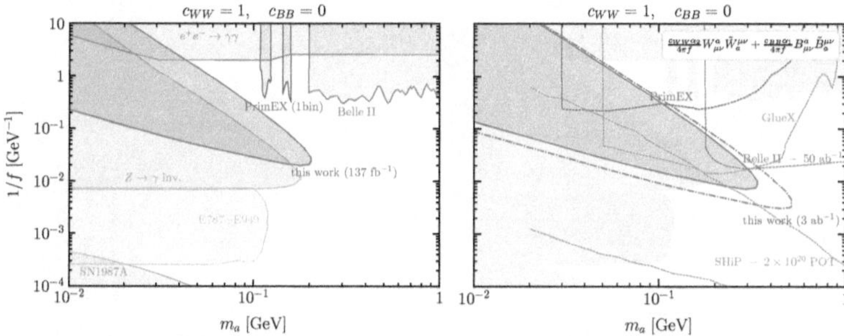

Figure 6.18: Constraints on ALPs coupled to W bosons. **Left:** Comparison of our current reach (red region) with existing constraints from star cooling constraints (green) [213], beam dump experiments (yellow) [213], Z invisible branching ratio (orange) [214, 215], limits on $e^+e^- \to \gamma\gamma$ from LEP data (violet) [214, 216–218], PrimeEX (purple) [219, 220], and Belle II (blue) [221]. **Right:** Projections of our constraints for a luminosity of $3ab^{-1}$ (red region). The solid and dot-dashed red contours correspond to the projections derived by using the same selections of the original CMS analysis, and the one derived by using a higher N_{hits} cut and assuming zero background. We compare our results with current constraints (gray shaded region) and projections for SHiP (orange) [183, 222], PrimEX (purple) [219, 220], GlueX (violet) [219, 223], , and Belle II (blue) [206, 222].

in the muon chambers selected as a signal by the analysis. Conversely, in the long lifetime area, the higher LLP multiplicity renders the limit more stringent than the case of Higgs decay to pairs of LLPs. Lowering the hidden confinement scale will increase the meson multiplicity inside hidden jets and therefore amplify this behavior.

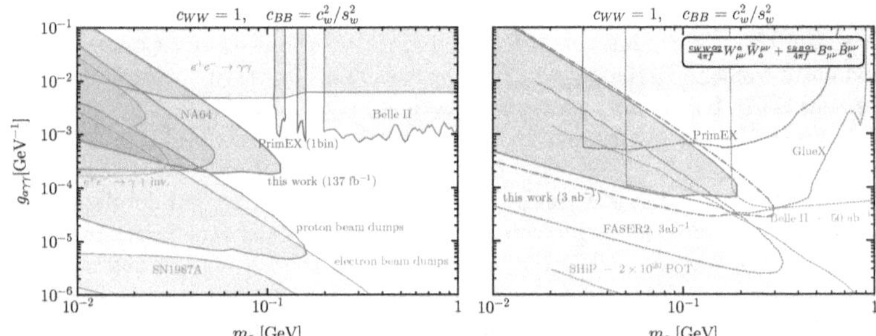

Figure 6.19: Constraints on ALPs coupled to electroweak gauge bosons, and with $c_{\gamma Z} = 0$. **Left**: Comparison of our current reach (red region) with existing constraints from star cooling constraints (green) [213], electron [200, 214, 224] and proton [188, 222, 225] beam dump experiments (pink and brown), limits from mono-photon searches at LEP (orange) [214, 226], NA64 (green) [201], PrimEX (purple) [219, 220], and Belle II (blue) [221]. **R**ight: Projections of our constraints for a luminosity of $3\,ab^{-1}$ (red region). The solid and dot-dashed red contours correspond to the projections derived by using the same selections of the original CMS analysis, and the one derived by using a higher N_{hits} cut and assuming zero background. We compare our results with current constraints (gray shaded region) and projections for FASER (brown) [212], SHiP (orange) [183], PrimEX (purple) [219, 220], GlueX (green) [219, 223], and Belle II (green) [206].

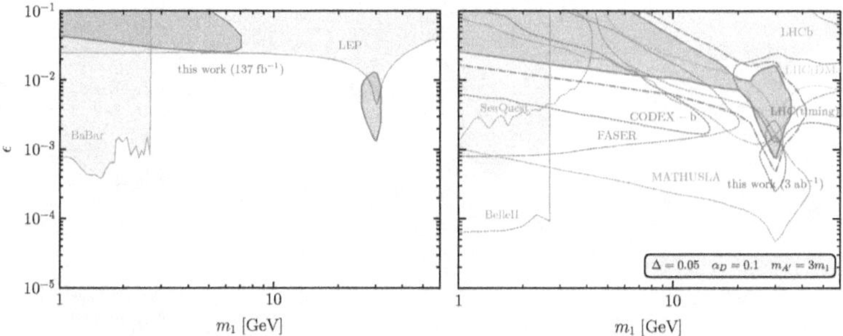

Figure 6.20: Constraints on inelastic DM models, assuming a normalized mass splitting of $\Delta = 0.05$, a dark coupling $\alpha_D = 0.1$, and mediator mass given by $m_{A'} = 3m_1$. **Left:** Comparison of our current reach(red region) with existing constraints from BaBar (green) [227, 228] and LEP (blue) [205, 227, 229, 230]. **Right:** Projections of our constraints for a luminosity of $3ab^{-1}$ (red region). The solid and dot-dashed red contours correspond to the projections derived by assuming the same selections of the original CMS analysis, and the one derived by using a higher N_{hits} cut and assuming zero background. We compare our results with current constraints (gray shaded region) and projections for BelleII (pink) [206], SeaQuest (gray) [203], FASER (blue) [231], MATHUSLA (orange) [232], CODEX-b (brown) [233], LHC (yellow and purple) [227, 234, 235], and LHCb (green) [227, 236].

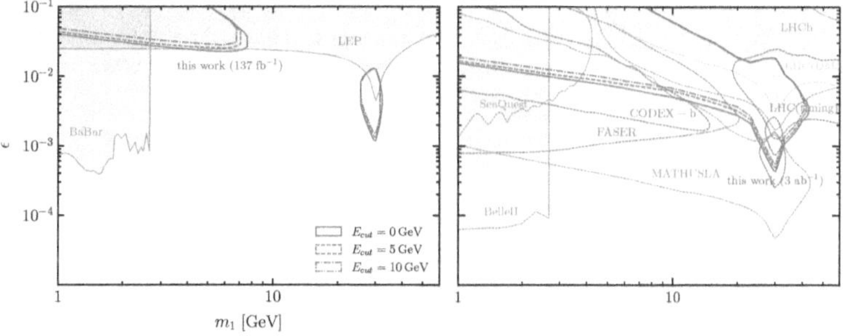

Figure 6.21: Constraints on the inelastic DM model for different choices of the lower cut on the LLP energy. The other constraints appearing in the plot are the same reported in Figure 6.20. The projections in the right panel are for a luminosity of $3ab^{-1}$ luminosity, and assuming a tighter N_{hits} cut and zero background.

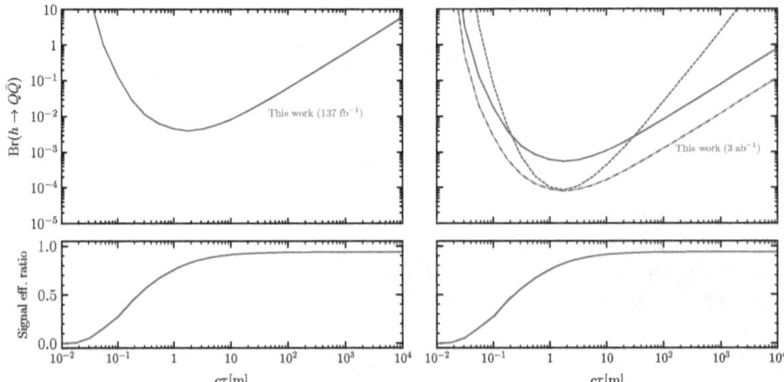

Figure 6.22: Current (upper left panel) and projected (upper right panel) constraints on the Higgs exotic decay into dark quarks of a confining hidden valley model. For the projections we report (solid, dashed, and dot-dashed lines) the results obtained by using the three search strategies discussed in the main text (rescaled CMS analysis, dedicated trigger, and higher N_{hits}).

6.6 Summary

In summary, it has been shown that this analysis proves very effective at constraining light LLPs, $m_{LLP} < O(\text{GeV})$, as long as they can be produced energetically in the LHC collisions and have $c\tau \lesssim O(\text{m})$. In fact, we found that the current version of such a search strategy not only provides a counterexample to the lore that LLP searches at ATLAS and CMS are limited at low masses by irreducible SM backgrounds, but it is already able to cover previously unconstrained parameter space in many models, competing with and complementing the reach of dedicated LLP detectors.

Through this work we have also discovered a few avenues of improvement for the CMS search, which was optimized for the twin Higgs signal model. As mentioned in Section 6.5, producing signal categories with lower MET requirements but in association with another object such as a photon, lepton(s) or b-jet, may improve limits on specific models such as ALPs and HNLs. Additionally, relaxing the cluster isolation requirement could also improve the sensitivity for many models, since in many cases the LLPs are produced inside (b-)jets. Examples include the case of a light scalar model, where S can be produced efficiently in b decays and would yield muon detector showers not isolated from a b-jet; the case of ALPs produced in b-flavored hadron decays or in hadronic showers via π^0-$\eta^{(\prime)}$ mixing; or the case of emerging jets [237] where showering within QCD and a Hidden Valley happens concurrently.

This reinterpretation work was made possible by the simplicity and reliability of the recasting provided by the publicly released information in HepData [143]. Finally, given the impact and ease-of-use of these additional information provided by CMS, we (the CMS collaboration) are actively working on providing the same information for the barrel muon detector showers that would allow future interpretation work. Additionally, it was found during the reinterpretation process that more finely spaced efficiency maps at low (E_{had}, E_{em}) to fully capture the turn-on shoulders, would be extremely helpful for reinterpreting models where LLPs with small visible energy are produced.

6.A   ATLAS Mono-jet Limit for Axion-like Particles Coupled to Gluons

This section summarizes the procedure used to reinterpret the ATLAS monojet limit on ALPs coupled to gluons [207]. The ATLAS collaboration already provides a lower limit on the ALP decay constant at a fixed ALP mass $m_a = 1\,\text{MeV}$ in this particular model and claims that such limit should hold for ALP masses up to approximately $1\,\text{GeV}$. This claim is motivated from ALP literature prior to the improved estimates on ALP lifetimes and branching ratios provided in Ref. [161] and it is modified in the region $0.1 - 1\,\text{GeV}$ due to the non-trivial behavior from ALP mixing with the neutral pseudoscalar mesons. To estimate the limit curve in this region we use our ALP+jet simulation to extract the 2D LLP energy and pseudorapidity distributions, convolve that with the lifetime model of [161], and require that the ALP does not decay in the ATLAS detector volume, for a fixed value of m_a and f. We then rescale the ATLAS limit for the ratio of the two efficiencies described above computed at $m_a = 1\,\text{MeV}$ and at a different mass point. This produces a function of $(m_a/1\,\text{MeV}, f/f_{\text{limit, 1\,MeV}})$. We then invert this function to solve for the limit on f as function of m_a as shown in Figure 6.17. As expected, the limit is fairly flat at low ALP masses but gets cut off earlier than $1\,\text{GeV}$ due to the ALP lifetime significantly changing after the m_η threshold. The steepness of the turn-off renders this limit curve a little sensitive to the specific geometric dimensions considered for the ATLAS detector.

The main limitation in the search for LLPs with muon detector using Run 2 data was the lack of dedicated trigger. As mentioned in the Chapter 5, the high p_T^{miss} trigger was used in Run 2, which had $\sim 1\%$ efficiency for the twin Higgs models.

For Run 3, a new dedicated L1 + HLT high multiplicity trigger (HMT) was successfully commissioned since 2022 using the CSCs that increases the trigger efficiency by at least a factor of 10. The new L1 seed is only implemented for CSCs due to the availability of four spare bits in the local trigger boards. There's not enough bandwidth in the DTs to add new L1 seed for Run 3, but for Phase 2 the entire DT L1 trigger system will be upgraded [238], so there will be new opportunities then to implement L1 DT shower seed.

The design and performance of the new HMT L1 seed and the corresponding HLT paths are described in Section 7.1 and Section 7.2, respectively.

7.1 New High Multiplicity Level-1 Seed

In the CSC local trigger, anode and cathode hits per bunch crossing were counted in each chamber (CSCs have both an anode-wire and cathode-strip readout). For an event to pass the trigger, we require that there exists at least one CSC chamber that has a large number of cathode hits and a large number of anode hits. Both cathode and anode hits are required because the rechit builder requires both types of information to reconstruct the muon detector showers in the HLT and in the offline reconstruction.

To evaluate the trigger rate of the HMT, we used zero bias data recorded in 2018 run era D. The trigger rate is calculated from the product of the fraction of passing data events multiplied by 30 MHz (40 MHz multiplied by the fraction of filled bunches). To evaluate the signal trigger efficiency, 23 dedicated twin Higgs signal samples with Run 3 run conditions were simulated, where the LLP decays to $b\bar{b}$, with Higgs boson masses between 125 and 1000 GeV, LLP masses between 1 and 450 GeV, and LLP proper decay lengths between 0.5 m and 100 m.

We determine the optimal thresholds for the anode and cathode hits for each chamber separately. In the CSC local trigger, the maximum number of comparator and wire hits are counted for each chamber in each ring separately: ME1/1, ME1/2, ME1/3, ME2/1, ME2/2, ME3/1, ME3/2, ME4/1, and ME4/2. The optimal thresholds for the anode and cathode hits are determined for each chamber separately. As an example, the distributions for the number of anode and cathode hits for ME2/2 are shown in Figure 7.1. It can be seen that the signal distributions have longer tails, while data distributions fall faster for both types of hits. In addition, the cathode hit distributions in data fall faster than the anode hit distributions, resulting in better sensitivity with cathode hits.

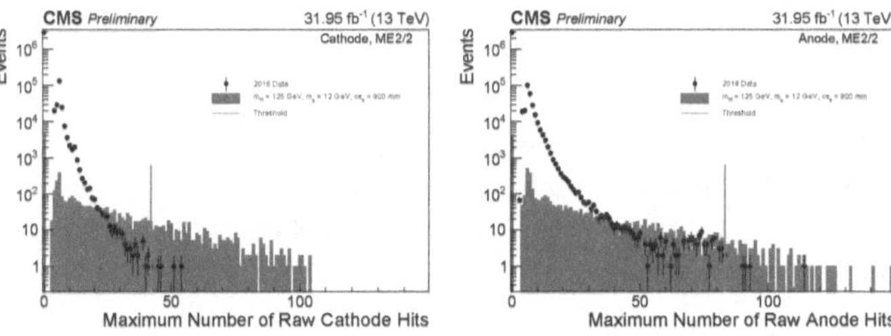

Figure 7.1: The N_{hits} distribution of the comparator digis (left) and wire digis (right) for signal and data in chamber ME2/2. The signal corresponds to m_{H} = 125 GeV, m_{S} = 12 GeV, and $c\tau$ = 900 mm.

Before determining the optimal thresholds for the number of anode and cathode hits, a few pre-selections on the hit time and minimum number of layers with hits are applied.

The cathode and anode hit time distributions for ME2/2 are shown in Figure 7.2. The anode hits from signal are concentrated in bunch crossing bin 8, corresponding to the nominal bunch crossing, while the anode hits from data are more evenly spread out. Therefore, we only count wire hits that are in the nominal anode bunch crossing, bin 8. The signal cathode hit time distributions are more spread out than that of the anode hits, but are still mainly concentrated in bins 6, 7, and 8. Bin 7 corresponds to the nominal bunch crossing in the cathode, while bin 6 and bin 8 correspond to the bunch crossing before and after, respectively. Placing a restriction on the comparator bunch crossing does not effect the efficiency or rate, but it makes the readout simpler. Therefore, we only consider cathode hits that are in the nominal ± one bunch crossing, i.e., bins 6-8.

In addition to selecting showers in specific time bins, the number of layers with hits in the chambers were also studied to help reduce the trigger rate. Figure 7.3 shows the number of layers distributions for cathode and anode hits in ME2/2. The anode distributions show that most signal showers produce hits in all six layers, but the

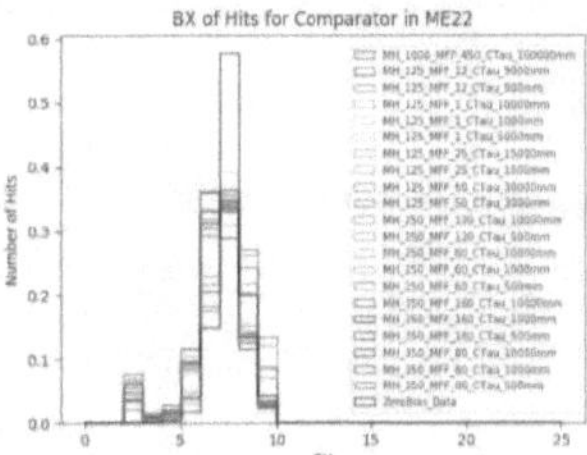 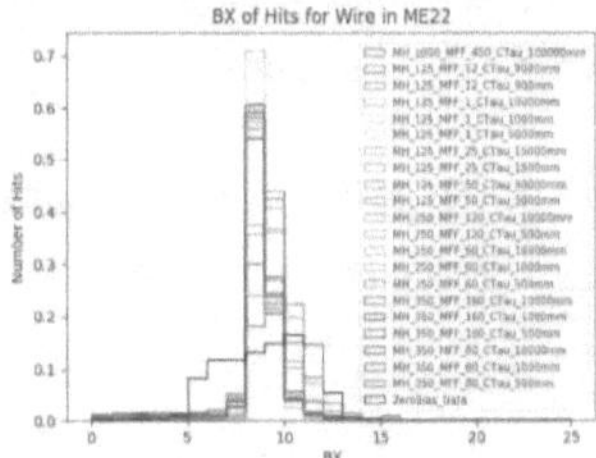

Figure 7.2: Timing distribution of the cathode digis (left) and wire digis (right) for signal and data in chamber ME2/2.

data has "showers" that only produce hits in a couple of layers, so we require at least 5 layers with hits to reject events with noisy anode layers. The requirement is > 95% efficient for signal MC, and provides a 30% rate reduction. The cathode distributions show that both signal and data mostly have more than 5 layers, and we require ≥ 5 layers for cathode hits as well to be consistent.

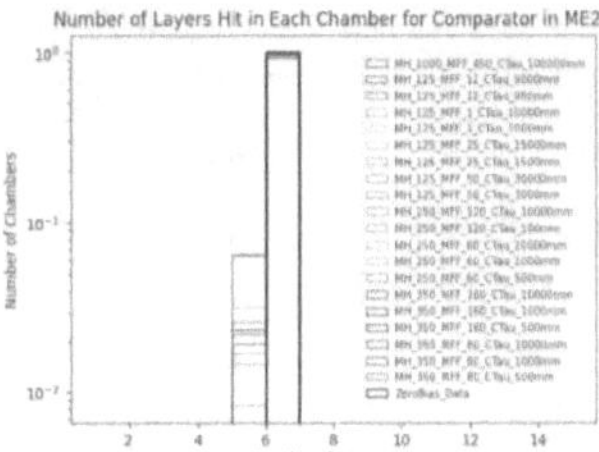 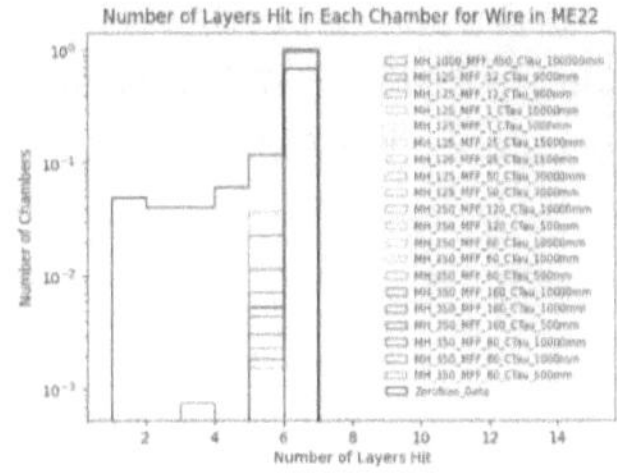

Figure 7.3: The number of layers distributions for cathode digis (left) and anode digis (right) for signal and data in Chamber ME2/2.

After implementing the bunch crossing and number of layers requirements, we optimize for the N_{hits} threshold to target an overall trigger rate of 1 kHz, while maximizing the signal efficiency. During commissioning, we have found that cathode counting for HMT is only possible for the inner rings that are quipped with the newer Optical Trigger Motherboard (OTMB). Therefore, for the outer rings showers are tagged with anode hits only, while for the inner rings showers are tagged with both cathode and anode hits. The final thresholds for the cathode and anode digis are shown in Table. 7.1. In general, there's more background in larger η region, so to control the trigger rate, a tighter working point is selected for the inner rings.

Overall, the rate study resulted in a rate of 0.64 kHz and signal efficiency 30% – 70% for different Higgs boson and LLP masses.

Table 7.1: Anode and cathode thresholds per chamber.

	ME1/1	ME1/2	ME1/3	ME2/1	ME2/2	ME3/1	ME3/2	ME4/1	ME4/2
Cathode	100	-	-	33	-	31	-	34	-
Anode	140	140	14	56	28	55	26	62	27

The logic structure of the HMT algorithm is shown in Figure 7.4. The logic for the local trigger is located on two boards: the Anode Local Charged Track Board (ALCT) and the (Optical) Trigger Motherboard (TMB) for each CSC chamber. The anode-shower provides accurate timing and the cathode-shower confirms that a shower occurred. In the ALCT we count the wire hits for each BX and determine if a shower exists. When a shower is seen in the ALCT, it will send the shower trigger data to the (O)TMB. The Cathode Local Charged Track logic on the newer OTMB will confirm the presence of a shower for the inner rings. The outer rings that are equipped with the older TMBs tag showers with only the anode hit information. Trigger data is then sent to the Muon Port Card (MPC), which does not contain any trigger logic for the HMT trigger. There are 60 MPCs, and each relays the LCT information from the 9 (O)TMB boards to the Endcap Muon Track Finder (EMTF) sector processors. The EMTF checks if at least one of the 45 chambers in a sector has a shower and sends the trigger data from each EMTF sector to the Global Muon Trigger (GMT). The GMT receives information from each of the 12 EMTFs, and performs a logical OR of all sectors to form a trigger data for the Global Trigger. The Global Trigger receives the trigger data from the GMT, which can then be used in signal trigger.

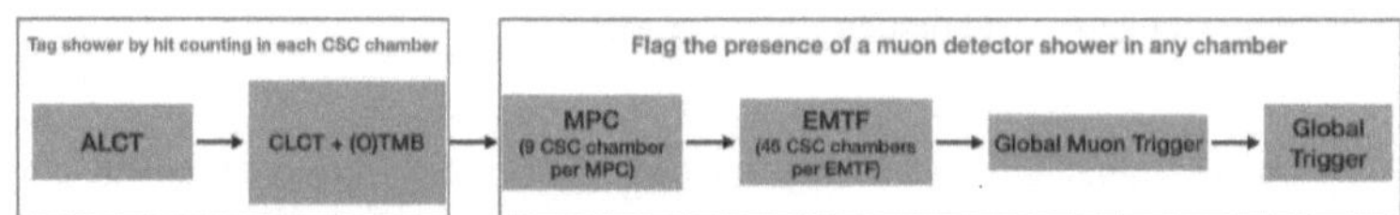

Figure 7.4: The L1 trigger logic of the HMT trigger.

The L1 rate, normalized by the number of colliding bunches, as a function of the number of pileup interactions is shown in Figures 7.5. The number of colliding bunches in this fill was 2448, which translates to L1 HMT trigger rate of 1.8 kHz at average pile-up of 50. The dependence of the normalized HLT rate on pileup is observed to be linear, as expected for single-object triggers.

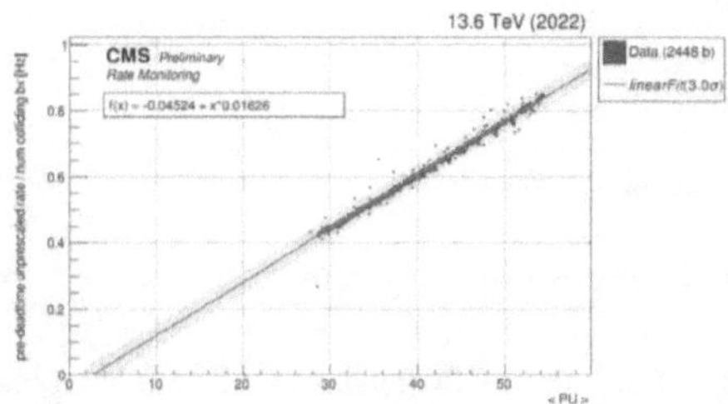

Figure 7.5: L1 rate of CSC High Multiplicity Trigger (HMT) per number of colliding bunches (bx) as a function of average pileup in an LHC fill [239]. The number of colliding bunches in this fill was 2448, which translates to L1 HMT trigger rate of 1.8 kHz at average pile-up of 50. The HMT trigger rate dependence on pile-up is extracted by using a linear fit. An uncertainty band corresponding to 99.7% of coverage (3σ) is also shown.

7.2 New High Multiplicity HLT Paths

In this section, we discuss HLT paths that are seeded with the L1 CSC hadronic shower trigger.

At the HLT level, the CSC and DT rechits are the same as the offline rechits. Therefore, we can perform the same rechit clustering to reconstruct muon detector showers and apply more sophisiticated selections that are more similar to the offline selections. The offline analysis uses DBSCAN algorithm, but to keep the integration to CMS software simple, we re-use the Cambridge-Aachen (CA) jet clustering algorithm to reconstruct MDS. We have studied the CA algorithm with different ΔR parameters and found that the algorithm with $\Delta R = 0.4$ maximizes the signal efficiency and gives similar efficiency as the DBSCAN algorithm. The CA algorithms with different ΔR parameters result in similar background rejection and thus trigger rates. Therefore, we use CA algorithm with $\Delta R = 0.4$ in the HLT.

After reconstructing the clusters, we designed two separate HLT paths, one targeting single CSC cluster per event and one targeting a CSC and a DT cluster in an event. For the single CSC cluster path, we apply similar selections as the Run 2 analysis to reject background to keep the HLT rate at around 1–10 Hz. The clusters are required to be in-time (-5 ns $< t_{\mathrm{cluster}} < 12.5$ ns) and clusters with hits in ME1/1 or ME1/2 are rejected. Then depending on the location and number of stations of the cluster we apply different N_{hits} threshold. The N_{hits} requirements are:

- $N_{\text{hits}} > 500$, if $|\eta| \geq 1.9$ and $N_{\text{stations}} = 1$,

- $N_{\text{hits}} > 200$, if $|\eta| < 1.9$ and $N_{\text{stations}} = 1$,

- $N_{\text{hits}} > 500$, if $|\eta| \geq 1.9$ and $N_{\text{stations}} > 1$, and

- $N_{\text{hits}} > 100$, if $|\eta| < 1.9$ and $N_{\text{stations}} > 1$,

Figure 7.6 shows that the single CSC cluster HMT path has a high L1 efficiency of close to 100% for large CSC clusters passing the HLT N_{hits} threshold.

For the DT-CSC cluster path, requiring an additional DT cluster in the event already rejects enough background to keep the HLT rate manageable, so we simply require an additional DT cluster with at least 50 hits that has no hits in MB1 on the L1-accepted events. The HLT rate is 7.6 Hz for the single CSC cluster path and 0.6 Hz for the DT-CSC cluster path in a particular run in October 2022 (Run 360019). Run 360019 had a peak instantaneous luminosity of $1.8 \times 10^{34} \text{cm}^{-2}\text{s}^{-1}$ with 54 peak pileup interactions. The overall trigger efficiency for the single CSC cluster path is 10-30% for different Higgs boson and LLP masses, and the efficiency for the DT-CSC cluster path is 10-20% for different Higgs boson and LLP masses. The efficiency is calculated with respect to the events that have at least one LLP decaying in the CSC for the single CSC cluster path and one LLP decaying in CSC and one decaying in DT for the DT-CSC path. For both HLT paths, the trigger efficiency is at least a factor 10 larger than the $\sim 1\%$ signal efficiency from using the $p_{\text{T}}^{\text{miss}}$-based triggers in Run 2. As of March 2024, the new trigger paths already recorded 23 and 27 fb^{-1} of data at $\sqrt{s} = 13.6\,\text{TeV}$ in 2022 and 2023, respectively. We are actively working on analyzing the 50 fb^{-1} of data recorded with the new trigger, which would give us potential to probe much smaller couplings than what was achieved in Run 2.

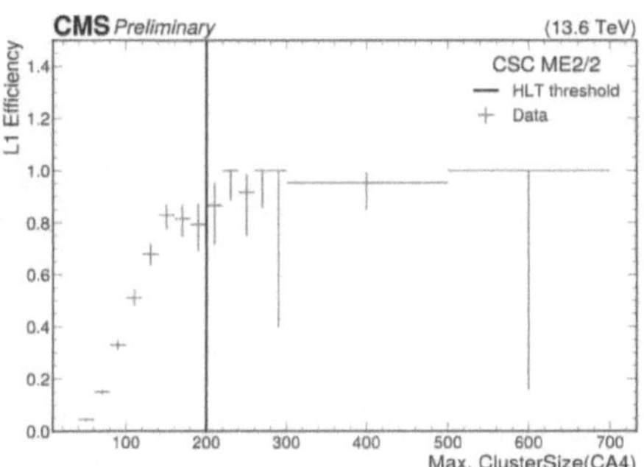

Figure 7.6: L1 efficiency of CSC High Multiplicity Trigger(HMT) as a function of the largest CSC rechit cluster size, which is a reconstructed quantity used at HLT [239]. The CA algorithm is used to cluster the CSC rechits with a distance parameter of $\Delta R = 0.4$. To evaluate the efficiency of individual CSC rings, the clusters position is restricted to be within a single CSC ring (e.g., ME2/2) and it is the only ring that contains more than 10 rechits in the cluster. The L1 efficiency is evaluated as the fraction of events in which the HMT is fired, given the above cluster selections are satisfied. A data sample of around 2M events triggered with zero bias triggers is used.

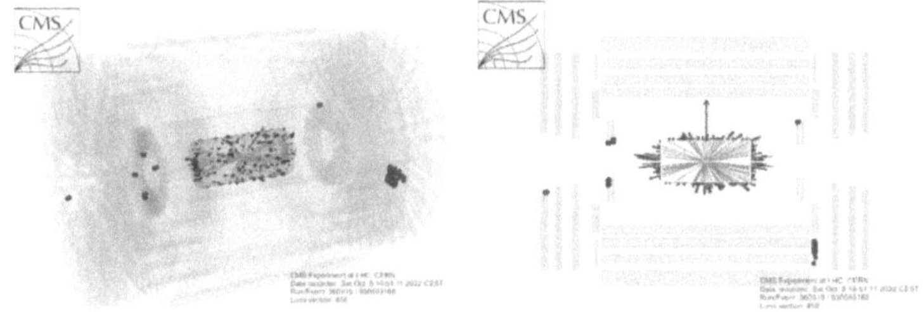

Figure 7.7: Event display of a collision triggered by the CSC HMT L1 and HLT [239]. CSC reconstructed hits are represented by blue dots in the muon end-cap region. This event features a CSC cluster of 210 hits in the ME1/3 ring. The event was recorded on October 8th, 2022.

Part III

Search for Dark Matter with QIS-enabled technology

Superconducting Nanowire Single Photon Detectors

8.1 Introduction

Diverse experiments and applications ranging from communications [240–242], metrology [243], remote sensing [244], and astronomy [245, 246] rely on single-photon detection. Since its first development in 2001 by Gol'tsman [247], SNSPD has become a leading detector technology for these applications, due to their high single-photon detection efficiencies exceeding 90% [248–250] low timing jitter of 3 ps [251, 252], and low dark counts of 10^{-5} Hz [253–255]. These properties have also popularized their use in several recent quantum-optics experiments, including quantum teleportation [256], quantum key distribution [254], characterization of quantum states [257–259], and quantum buffer memories [260, 261]. Detailed surveys of applications can be found in Ref. [262, 263]. The diverse range of applications of SNSPDs has raised the need to make such devices available commercially. Today, SNSPDs are supplied commercially by companies such as SCONTEL, Single Quantum, PhotonSpot, Quantum Opus, and Photech.

However, the application of SNSPDs has been limited in high energy physics (HEP), where other single photon detectors such as single-photon avalanche diodes (SPADs), photomultiplier tubes (PMTs), silicon photomultipliers (SiPMs) are often used. Even though SNSPDs offer unrivaled detection metrics with its unprecedented combination of high detection efficiency with high time-resolution and low noise that remain challenging in SPADs, PMTs, and SiPMs, the limited application of SNSPDs in HEP experiments is mostly due to its small active area ($\sim 100 \, \mu m^2$). In recent years, the active areas of SNSPDs have steadily increased from the scale of a single-mode optical fiber diameter ($\sim 100 \, \mu m^2$) towards the mm^2 regime [264–267], enabling their recent applications in high energy physics such as DM direct detection experiments. The first ever HEP experiments to use SNSPDs are direct DM searches that used SNSPDs as the target material for DM absorption [255] and DM-electron scattering [268].

In this part of the book, I will present two new complementary experimental techniques to directly detect DM with SNSPSDs coupled to: i) a novel broadband focusing reflector or ii) a bright cryogenic scintillator, gallium arsenide (GaAs), that can significantly enhance the sensitivity to DM thanks to the unique capabilities of SNSPDs. This chapter presents an overview of SNSPDs. The two novel applications in DM searches are detailed in Chapters 9 and 10, respectively.

The rest of this chapter is organized as follows. The detection mechanism of SNSPDs is introduced in Section 8.2. Section 8.3 describes the properties that are important for the SNSPDs: detection efficiency, dark counts, photon wavelength sensitivity, and timing jitter. Finally, the summary is presented in Section 8.4.

8.2 Detection Mechanism

At its core, an SNSPD is a current-carrying superconducting nanowire meandered on a substrate. The nanowires typically have widths ranging from tens of nm to a few μm and lengths ranging from hundreds of μm to several mm. The nanowires are more commonly made of low-gap amorphous superconductors such as molybdenum silicide (MoSi) and tungsten silicide (WSi) [269], but they have also been made out of dozens of other superconductors, depending on the applications. To operate the SNSPDs, they are usually current-biased and operated at a temperature of 1-4 K depending on the critical temperature of the superconducting material. Current-biasing of the SNSPD is achieved through the use of a bias-tee, and the readout portion of the tee is coupled to an amplifier and counting electronics that are used to detect the single-photon events and register the corresponding voltage pulses.

The basic detection process of an SNSPD is shown in Figure 8.1. To allow for quantitive modeling and design optimization, the detection process is divided into five steps: The device begins in a superconducting, current-carrying state, until (1) a photon is absorbed and (2) generates a hotspot of excited quasiparticles and phonons through a process called downconversion. The evolution of the quasiparticles leads to an instability of the superconducting state and results in the formation of vortices. As the vortices move due to the current flow, they dissipate energy, leading to (3) the formation of a normal domain across the entire cross-section of the nanowire. Subsequently, (4) the normal domain of the nanowire grows along the length of the wire due to internal Joule heating. The increasing resistance, on the order of kΩ, diverts the bias current from the nanowire to the readout electronics. Once current is diverted from the nanowire, (5) the nanowire cools down and recovers to the superconducting state, and the bias current returns to the device.

The initial absorption of a single photon within the active detector area is well described by a classical electromagnetic theory. This allows for the use of established modeling tools [271] to optimize optical absorption in the superconducting layer for a desired wavelength range.

The absorption of a photon usually generates an electron-hole pair. This excitation

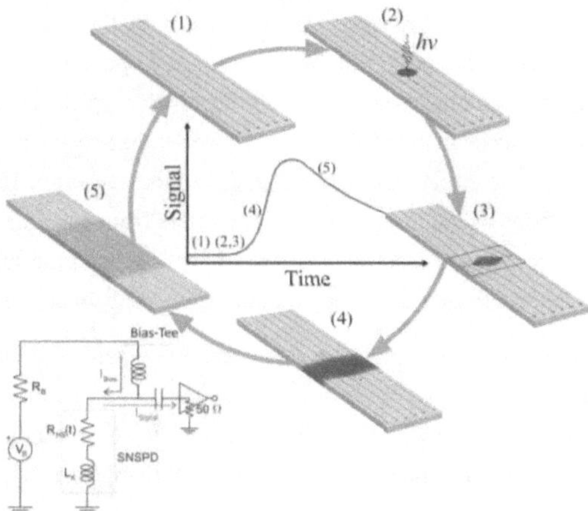

Figure 8.1: Schematic of the SNSPD detection process (top right), divided into five steps: photon absorption, generation of hotspot of quasiparticles and phonons, emergence of normal domain in the nanowire, re-direction of bias current to readout electronics, and detector recovery. An SNSPD is typically biased through a bias-tee with a DC port carrying the bias current and the RF port coupled to a low-noise amplifier (bottom left). The figure is adapted from [270].

quickly interacts with the electronic and lattice system to generate a thermalized hotspot in a process called downconversion. The electron and hole first interact with the electronic system through high momentum transfer collisions, generating a small number of high-energy electrons and holes. Once the energy of these excitations approaches a few times the Debye energy, phonon interactions dominate and most of the energy is transferred to the phonon system. After a few picoseconds [272], the electron and phonon systems thermalize and this excitation region is typically referred to as a hotspot [247]. The downconversion process can be modeled through deterministic kinetic equations for electrons and phonons [272] or through a stochastic loss of excitation energy into the substrate [273]. Combining the ideas from the deterministic and stochastic models can describe a complete set of measurements qualitatively [270, 274], but these existing models require further developments to be able to fully describe the physical processes quantitatively.

As the number of quasiparticles increases, an instability of the superconducting state occurs. Models of this process have been refined over the years [275], but the current understanding [272] is that the suppression of superconductivity is led by the formation of vortices in the nanowire. The nature of this process depends on a number of parameters, including the bias current, the operating temperature, the width of the nanowire, and the location of the hotspot. For typical geometries, if the hotspot is located in the center of the nanowire, a vortex-antivortex pair forms, with each moving in opposite directions toward the edge of the nanowire. If the photon is absorbed near the edge of the nanowire, a vortex enters from the edge and traverses the width of the wire. In both cases, the motion of vortices heats the superconductor and contributes to the destruction of superconductivity in the nanowire.

Subsequently, once a normal domain reaches across the nanowire width, it grows along its length due to internal Joule heating. Joule heating continues as long as there is current flowing through the normal domain. However, the normal domain has a large impedance. The sheet resistance for SNSPD materials typically ranges from a few hundred to a thousand ohms per square and the normal domain typically has a length of a few to a few tens of squares. The large impedance of the normal domain redirects the bias current from the nanowire to the readout electronics. Once the current is diverted, the resistive area starts to cool down and return back to the superconducting state, which then invites the bias current back to the nanowire.

The detection process depends strongly on the material properties and the device geometry. The community currently does not have an accurate means of predicting the detector performance for an arbitrary set of material properties and device design. While models for steps 1, 4, and 5 can be used to predict and develop successful designs, models for steps 2 and 3 are missing such capabilities and are open problems and challenges for future development.

8.3 SNSPD Properties

8.3.1 Detection Efficiency

The detection efficiency of the SNSPD is the probability of detecting a photon, which is the product of the absorption efficiency and the internal detection efficiency. The absorption efficiency is the probability of a photon that arrives at a device to be absorbed by the wire. This efficiency depends not only on the material absorptance, but also the contribution of additional design factors that could improve the absorption efficiency, such as mirrors, fill factor, cavities, and antenna. The absorption efficiency depends on the photon wavelength, incoming angle, and polarization.

The internal detection efficiency is an intrinsic SNSPD properties that defines the probability of a photon absorption event to suppress the superconductivity to result in a measurable signal. The internal detection efficiency is usually characterized by a long-range saturation of the efficiency as a function of the bias current. Typically, SNSPDs based on amorphous materials (tungsten silicide and molybdenum-based) show higher efficiency than the crystalline SNSPDs. The highest system detection efficiency that has been reported to date is 98% on 30-50 μm diameter MoSi SNSPD [248] for 1550 nm photons.

8.3.2 Dark Count

Dark counts or false counts are any measurable voltage pulses on the SNSPDs that do not originate from a photon that was sent to the device intentionally either from a laser or a dark matter candidate. Such dark counts arise either from straying photons or fluctuations from the superconducting to normal state in a device that is held close to the phase transition. Thermal fluctuations [276], fluctuations in the bias currents, quantum fluctuations in the amplitude [277] and phase [278] of the superconducting order parameter, and fluctuations in the electronics can all contribute to the total dark counts. Bartolf et al. [276] suggested that the intrinsic dark counts that are not from straying photons come from three competing mechanisms: i) single-vortex motion, ii) vortexantivrotex unbinding, and iii) vortex hopping. The dominancy of each of these mechanisms depends on experimental parameters.

Most solutions to decrease the dark count rate (DCR) involve tuning extrinsic parameters to operate at a state that is far from the phase transition, such as lowering the operating temperature and lowering the bias current. Additionally, better thermal and optical isolation also help removing background from straying photons, as will be demonstrated in the measurements presented in Chapter 9. However, a few design and material considerations can also decrease the DCR. The demand for low sensitivity to fluctuations dictates that the material should be homogeneous and should have a constriction-free and current-crowding free design. The lowest DCR that has been reported to date is 6×10^{-6} count per second (cps) on a 0.4×0.4 mm^2 WSi SNSPD [255].

8.3.3 Photon Wavelength Sensitivity

A major advantage of SNSPDs over competing technologies is their sensitivity to photons with energy lower than that of visible light, mainly in the near-infrared region, at the communication-relevant wavelengths. For a given device, photons with shorter wavelength or higher energy are detected more efficiently [279–281].

This is due to the fact that higher input energy breaks more Cooper pairs, give rise to a larger hotspot, and thus a higher detection efficiency. The detection energy threshold of an SNSPD scales with the characteristic energy [272]:

$$E_0 = 4N(0)(k_b T_c)^2 V_0 \tag{8.1}$$

where $N(0) = (2\rho e^2 D)^{-1}$ is the density of states per spin at the Fermi level in the normal state, k_b is Boltzmann's constant, T_c is the critical temperature, V_0 is the characteristic volume, D is the diffusion coefficient, ρ is the resistivity of the film, and e is the electron charge.

For a WSi superconducting film, the detection threshold can be reduced by increasing the silicon content which increases the resistivity of the film, reducing the density of states and T_c of the material. Increasing the silicon content also reduces the T_c of the material. This conclusion is supported, for example, by the demonstration of saturated detection efficiency for WSi SNSPDs with photon wavelengths up to 29 μm [282].

8.3.4 Timing Jitter

The excellent timing resolution of the SNSPD is one its most attractive property against competing technologies. The timing resolution of the SNSPDs, commonly referred to as jitter, is usually quantified by the width of the temporal distribution of the SNSPD output signals with respect to the photon arrival time. The ability to directly observe the intrinsic timing jitter has also been difficult, since the jitter measurements are often limited by instrumental temporal uncertainties from the readout electronics and the single photon source. The jitter from electronic noise can often be reduced by engineering devices with higher bias current, faster rise time and using lower noise cryogenic amplifiers [283].

Considerable effort has been made in the SNSPD community to understand the fundamental limit of the timing jitter, but the mechanisms that dictate the intrinsic timing jitter are still not completely understood. Ref. [283] suggested that the fundamental limit corresponds to the time it takes for a vortex to cross the width of an SNSPD, which is about 1 ps. Likewise, others have suggested that for amorphous superconductors, the fundamental timing jitter is dictated by Fano fluctuations, which is also about 1 ps. Finally, the Berggren group [284] related the fundamental limit to the specific position that the photon was absorbed along the nanowire.

Empirical studies have suggested that higher bias current [251, 285], critical temperature [285], and photon energy [251] can improve the timing jitter. It has been found that crystalline materials such as niobium nitride have exhibited smaller timing jitter than amorphous materials (WSi, MoSi) [251], which could be attributed to the larger energy gap and critical temperature that such materials usually exhibit The best timing jitter that has been reported to date is 2.5 ps for visible wavelengths and 4.3 ps at 1550 nm on a niobium nitride SNSPD [251].

8.4 Summary

Since its first development in 2001, SNSPD has become a leading detector technology for many applications, due to their high single-photon detection efficiencies, low timing jitter, and low dark counts. However, even though single-photon detectors are commonly used in many HEP experiments, SNSPDs haven't found its applications in HEP until recently, due to their limited active areas. The demonstrations above only achieved a single high performance metric with a given device, so additional developments are needed to demonstrate high efficiency, low dark counts, and high timing resolution simultaneously. Additionally, the high performance metrics were demonstrated for small area SNSPDs. Recent advancements that allowed the fabrication of mm^2 SNSPDs enables their applications in HEP, such as direct DM detection experiments. Chapters 9 and 10 will present the first efforts to characterize these large area (mm^2) SNSPDs and their applications in DM detection experiments.

Search for Axion with SNSPDs at the BREAD Experiment

9.1 Introduction

In this chapter, we will discuss another missing piece of our current understanding of particle physics, known as the strong CP problem: why is there no CP violation in the strong force? It turns out that the axions, that were originally proposed as a solution to the strong CP problem, can also serve as an excellent dark matter candidate. There has been many efforts to search for axions, since its first proposal in 1977, but no axions have been found yet. The most-sensitive axion detection experiments today are based on radio-frequency resonant-cavity haloscopes. However, this strategy is only able to search for light axions on the μeV scale and has the disadvantage of narrow-band tuning to the axion mass, which is an unknown parameter. Therefore, the Broadband Reflector Experiment for Axion Detection (BREAD experiment) proposes a novel experimental design that can search for decades of axion masses from μeV to eV without tuning. The experiment uses a novel cylindrical dish resonator that fits in a solenoid, allowing axions or dark photons to convert to photons, regardless of their masses. The axion- or dark photon-converted photons are then focused by a parabolic focusing reflector to a low noise photon detector. A first stage pilot dark photon experiment using SNSPDs is being planned for at Fermilab and ongoing work is being carried out to characterize the SNSPDs to commission the pilot experiment.

This chapter is organized as follows. The axion as a solution to the strong CP problem, axion models, and axion detection techniques are introduced in Section 9.2. In Section 9.3, the conceptual design and sensitivity estimate of the BREAD experiment is described in detail. The work that I have done to characterize SNSPD towards a pilot dark photon experiment is detailed in Section 9.4. Finally, the summary and outlook are presented in Section 9.5.

9.2 The Axion

9.2.1 The Strong CP Problem

Quantum chromodynamics (QCD) is a theory that describes the strong interactions of colored quarks and gluons. The QCD Lagrangian can be written as:

$$\mathcal{L}_{\text{QCD}} = -\frac{1}{4}G^a_{\mu\nu}\tilde{G}^{a\mu\nu} + \sum_f \bar{\psi}_f(i\gamma^\mu D_\mu - m_f)\psi_f \tag{9.1}$$

where $G_{\mu\nu}$ is the gluon field-strength tensor, f labels the quark flavor, ψ_f is the quark field, γ^μ is the gamma matrix, D_μ is the covariant derivative, and m_f is the quark mass.

The non-Abelian nature of the gluon field results in degenerate topologically distinct vacuum states [286], with each vacuum state in a distinct class labeled by the topological winding number, n. The physical vacuum state, which is referred to as the "θ-vacuum", is a superposition of these states:

$$|\theta\rangle = \sum_n e^{in\theta} |n\rangle \tag{9.2}$$

where θ is an arbitrary parameter that needs to be measured. The θ that gives rise to the QCD vacuum state and another source term that comes from the quarks' Yukawa coupling with the Higgs allow for the presence of an effective interaction that violates charge and parity (CP):

$$\mathcal{L}_\theta = \frac{\bar{\theta}}{32\pi^2} G_{\mu\nu}^a \tilde{G}^{a\mu\nu} \tag{9.3}$$

where $\bar{\theta} = \theta + \arg(\det(M))$ is the CP-violating phase and M is the quark mass matrix. θ and $\arg(\det(M))$ are unrelated angles that are not fixed by the theory, so their sum $\bar{\theta}$ is also an arbitrary angle that needs to be determined experimentally and the size of which determines the amount of CP violation in QCD.

The most precise test of $\bar{\theta}$ is the measurement of neutron electric dipole moment (EDM). The existence of neutron EDM is inherently CP-violating, as illustrated in Figure 9.1. The neutron's EDM is proportional to $\bar{\theta}$ and can be written as [287]:

$$d_N \sim \bar{\theta} \frac{em_q}{m_N^2} \tag{9.4}$$

$$\sim 10^{-16} \bar{\theta} \text{ e-cm} \tag{9.5}$$

where e is the electron charge, m_N is the neutron mass, and m_q is an effective quark mass [288] often written as $m_q \sim \frac{m_u m_d}{m_u + m_d}$. The latest experimental effort to measure the neutral EDM has instead set a stringent upper limit on the value [289]:$d_N < 1.8 \times 10^{-26}$e-cm, resulting in a limit on the CP-violating phase: $\bar{\theta} \lesssim 10^{-10}$. If one were to pick a random angle, one would expect $\bar{\theta}$ to be $O(1)$, but the upper limit set by the neutron EDM measurement is clearly much smaller. Additionally, there is no physics reason for two unrelated quantities θ and $\arg(\det(M))$ to cancel each other. This fine-tuning problem of $\bar{\theta}$ is referred to as the "strong CP problem."

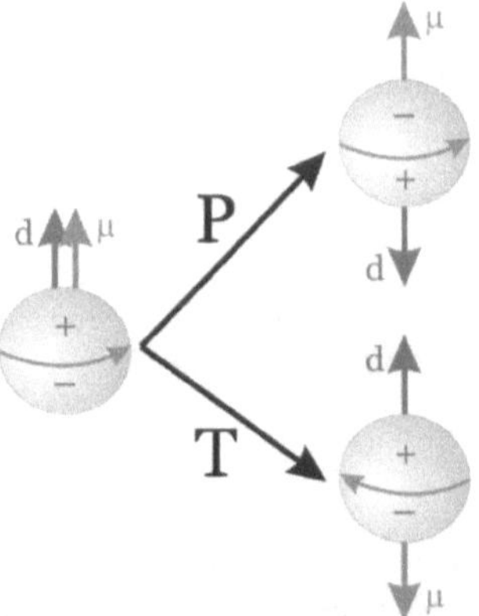

Figure 9.1: Illustration of a neutron under T (CP) and P transformation. The associated charge (red + and −), spin (blue arrow), magnetic dipole moment (μ), and electric dipole moment (d) are shown. Applying time transformation preserves the charge distribution, but reverses the spin. Therefore, the directions of the electric and magnetic dipole moment that are measured with respect to the spin also change direction, violating time-reversal symmetry. Based on the CPT theorem, a nonzero neutron electric dipole moment would result in CP violation. Figure from Andreas Knecht.

9.2.2 The Peccei-Quinn Solution

Roberto Peccei and Helen Quinn proposed a solution to the strong CP problem in 1977 [290, 291]. The Peccei-Quinn solution promotes θ from being a static number to a dynamic field by adding a new global $U(1)$ chiral symmetry, the Peccei-Quinn symmetry $U(1)_{\mathrm{PQ}}$, that spontaneously breaks at a large energy scale, f_a. Additionally, the PQ solution also requires at least one quark that gains mass by interacting with the complex scalar field via a Yukawa interaction.

In the early formation of the universe, when the universe temperature is above the symmetry breaking scale ($T > f_a$), the symmetry is unbroken and the minimum of the potential is at zero. As the temperature cools ($T < f_a$), the potential changes to the classical "wine bottle" or "mexican hat" form:

$$V(\phi) = |\phi^2| - \frac{f_a^2}{2} \tag{9.6}$$

where f_a is the symmetry-breaking scale above which CP is violated. To remain in a minimum energy state, the $U(1)_{\mathrm{PQ}}$ symmetry is spontaneously broken and the non-zero expectation value causes the quark to generate mass, $m_q \sim \langle\phi\rangle$. When the temperature further cools to $T < \Lambda_{\mathrm{QCD}}$, interactions between the field and the gluon causes the entire potential to tip, creating a preferred value of $\theta = \bar{\theta}$ that is unrelated to the initial configuration $\theta = \theta_i$ when the symmetry spontaneously broke. The stages of the symmetry breaking are shown in Figure 9.2.

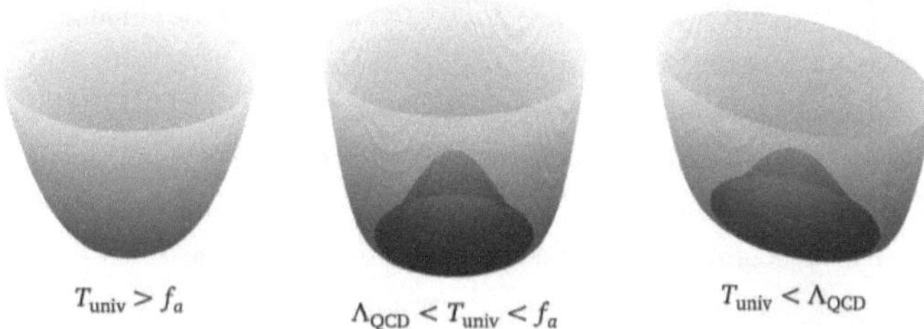

Figure 9.2: The stages of Peccei-Quinn symmetry breaking. Left: In the early, hot universe ($T_{\mathrm{univ}} > f_a$), the symmetry is unbroken with the minimum of the potential at zero. Middle: When the universe cools below the symmetry breaking scale, the shape of the potential changes to the classical "wine bottle" with energy degeneracies at different azimuthal angles. The symmetry is spontaneously broken and gains a vacuum expectation value. Right: When the universe cools to $T_{\mathrm{univ}} < \Lambda_{\mathrm{QCD}}$, the gluon interacts with the field and introduces a tip that breaks the energy degenracies, creating a preferred minimum of the potential. Figure reprinted from [292].

The interactions that lead to the potential tipping add a third contribution to the CP-violationg Lagrangian term $\mathcal{L}_\theta$, so the new term is now written as:

$$\mathcal{L}_\theta = \frac{1}{32\pi^2}(\bar{\theta} + \frac{a}{f_a})G^a_{\mu\nu}\tilde{G}^{a\mu\nu} \tag{9.7}$$

To minimize the total energy of the state, requires the coefficient $\bar{\theta} + \frac{a}{f_a}$ to be 0. Physically, this represents the tipping of the potential driving θ to $-\frac{\langle a\rangle}{f_a}$. Thus, $\mathcal{L}_\theta$ approaches zero, solves strong CP problem.

9.2.3 Axion Models

Shortly after the proposal of the PQ solution in 1977, Steven Weinberg and Frank Wilczek independently realized that the spontaneous breaking of the PQ symmetry and its explicit breaking from the chiral anomaly would imply the existence of a massive pseudo-Nambu Goldstone boson, which was named the "axion" [293, 294]. The mass of the axion (m_a) can be calculated by expanding the tipped potential around the minimum [295]:

$$m_a^2 = 5.681(51) \left(\frac{10^9 \, \text{GeV}}{f_a} \right) \text{meV}. \tag{9.8}$$

9.2.3.1 PQWW Model

The original work by Peccei, Quinn, Wilczek, and Weinberg made minimal additions to the SM and became known as the PQWW model [290, 291, 293, 294]. They assumed that the quark that is responsible for tipping the potential is a SM quark and that all of the SM fermions are PQ-charged. Additionally, the PQ symmetry breaking occurs at the electroweak scale. The relatively low f_a results in a strong axion coupling that makes the PQWW model immediately testable experimentally, through heavy meson decays, reactor emissions, and neutrino beam dump experiments [294]. However, no axion signatures were found experimentally, which led to the theories with much larger symmetry breaking scale, thus more weakly-coupled axions called "invisible" axions.

9.2.3.2 Invisible Axions

There are a number of theoretical models that make axions "invisible" to experiments, by having PQ symmetry breaking scale much larger than the electroweak scale. There are two benchmark models that are widely accepted for experimental searches: the Kim-Shifman-Vainshtein-Zakharov (KSVZ) axion [296, 297] that couples only to quarks and Dine-Fischler-Srednicki-Zhitnitsky (DFSZ) axion [298, 299] that couples to both quarks and leptons. There are also a number of other models that are either variations of these two models or take on entirely different approaches. However, the axion couplings vary only by $O(1)$ factors from the different theories. Therefore, experimentalists build detectors to probe axion couplings that can reach the "QCD axion band," which is the parameter space covered by these models that can solve the strong CP problem in QCD.

The most commonly explored interaction, as is the focus of this book, is the axion-photon interaction. The size of the axion to two photon coupling is given by [300]:

$$g_{a\gamma\gamma} = \frac{\alpha}{2\pi} \frac{C_{a\gamma}}{f_a} \approx 2.0 \times 10^{-14} \, \text{GeV}^{-1} \frac{m_a}{100 \mu eV} C_{a\gamma} \tag{9.9}$$

where $C_{a\gamma} = \frac{E}{N} - 1.92(4)$ is a model-dependent constant that is a function of the electromagnetic (E) and color (N) of the axial symmetry. $C_{a\gamma}$ is usually assumed to be of order one. For the KSVZ and DFSZ axion models, $|C_{a\gamma}|$ are ~ 1.9 and 0.7, respectively, corresponding to E/N values of 0 and 8/3, respectively.

9.2.3.3 Axion-like Particles & Dark Photons

A more general class of particles, axion-like particles (ALPs), that have independent coupling and mass, are often considered as well. They do not necessarily lie on the QCD band, so they do not solve the strong CP problem. However, they are mentioned here, because many experiments that aim to search for QCD axions naturally are also sensitive to ALPs. ALPs are predicted to arise more generically in low-energy effective field theories emerging from string theory [301–307].

Additionally, many axion's experimental signatures are also shared by another DM candidate, called the dark photons. The dark photon is the gauge boson of a new dark $U(1)$ symmetry added to the SM gauge group, under which the SM fields are uncharged. It shares a number of phenomenological features with the axions, notably the possibility of converting to SM photons [308, 309]. The main practical difference between the two, as will be discussed in more detail in Section 9.2.5, is that the axion-photon conversion requires an external magnetic field, while dark photon-photon mixing is an inherent feature of the model. While the dark photons do not solve the strong CP problem in QCD, they are viable cold DM candidates [310, 311]. Many production mechanisms have been proposed to generate a sufficient abundance of them in the early universe [312–321].

9.2.4 Axion Cosmology

Due to the high energy scale of the PQ symmetry breaking, axions are produced in the very early universe. Knowledge of the early universe can provide insight into present-day axion phenomena, and if axions are discovered, their properties may provide a window into the history of our universe that otherwise cannot be probed. The production mechanisms and behavior of axions will be described in this section in a few different cosmological scenarios. We will see that the axions are excellent DM candidates, since they can be produced in the correct abundance, and are weakly-coupled, cold, and stable.

The origin of the axion particle was discussed in Section 9.2.2. We have mentioned that the QCD interaction of a complex scalar tips its potential. The field, a, associated with the angular freedom of the complex scalar field would then obtain a non-constant potential, with a minimum at $\bar{\theta}$. When the potential gains a minimum, the field already has an initial value at θ_i, when the PQ symmetry was initially spontaneously broken. The fact that $\theta_i \neq \bar{\theta}$ leads to the production of cold axions, and this is called the misalignment mechanism. This population of cold axions can make up the cold dark matter in the universe.

The equation of motion of the axions "rolling" from the θ_i to $\bar{\theta}$ can be written as:

$$\ddot{a} + 3H(t)\dot{a} + \frac{\partial V}{\partial a} = 0 \tag{9.10}$$

where $H(t)$ is the Hubble constant and V is the potential after tipping. The equation is similar to that of a damped oscillator in an expanding universe. If we use the quadratic approximation near the minimum of the potential, $V(a) \approx \frac{1}{2}m_a^2 a^2$, then the equation of motion simplifies to

$$\ddot{a} + 3H(t)\dot{a} + m_a^2 a = 0. \tag{9.11}$$

Thus, $3H(t)$ corresponds to the damping and m_a^2 corresponds to the oscillation frequency. Since $H(t)$ is temperature-dependent and evolves as the universe cools, the axion will experience different types of oscillations (over-damped, under-damped, or undamped oscillations) as the universe expands and cools. The changing behavior of axion as a function of $1/T$ is shown in Figure 9.3. As shown in the figure, the axion is initially stuck at θ_i, and then starts to slowly roll down the potential, and eventually oscillates around the minimum with decreasing damping, until it reaches a state with stable oscillation. The final form of the axion field is a stable oscillation with the form of

$$a(t) = C \sin(m_a t) \tag{9.12}$$

where C is the the amplitude of the oscillation. If the axions were to account for all of the present day local dark matter density, then $C \approx \sqrt{2\rho_{\mathrm{DM}}}/m_a$, where ρ_{DM} is the local dark matter density. As the oscillations begin when $m_a \approx 3H$, the axion starts to form into pressureless massive particles. After that, the axion particles would then dilute in the expanding universe just like visible matter and fall into gravitational potentials of clusters of matter to form the early structure of the universe and eventually become the dark matter found in galaxies.

The relic energy density of the axion depends on the axion mass (inversely proportional to f_a) and the misalignment angle between the initial field value and the potential minimum. The latter sets the axion energy density at the moment of PQ

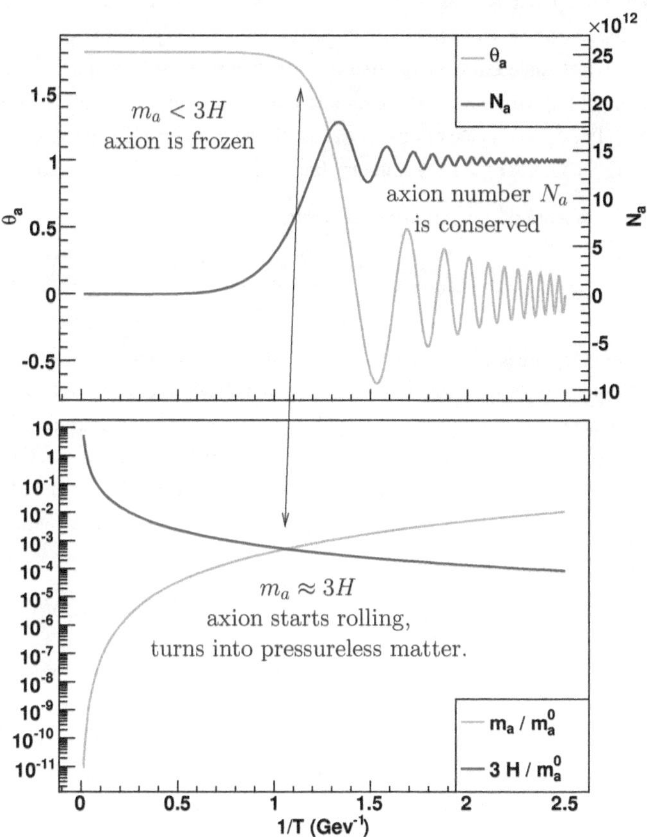

Figure 9.3: Evolution of the axion field, reprinted from [322]. Top: Qualitatively the changing behavior of axion over time, as the universe cools and expands. The axion is first frozen at its misalignment angle, then begins to slowly roll down to the minimum of the potential, and eventually oscillates around the minimum with decreasing damping. N_a is the number of axion particles per comoving volume, demonstrating that as axion starts rolling, it turns into pressureless matter. Bottom: The behavior of the Hubble constant H and axion mass m_a over time are shown. Both H and m_a are dependent on temperature, so their behavior also depends on the specific cosmological model.

symmetry breaking, while the former determines how diluted the energy density gets. The timing that oscillation starts is driven the by the relative sizes of H and m_a, a more massive axion leads to an earlier transition to axion particle, longer time for dilution, so a smaller axion energy density. Every non-casually connected region of space will end up with an independent initial angle θ_i at the time of PQ symmetry breaking. The consequence of this range of different initial angles are very different depending on whether the PQ symmetry breaking happens before or after inflation.

The energy density of the axions can be written as [323]:

$$\Omega_a \sim 0.15\theta_i^2 \left(\frac{f_a}{10^{12}\,\mathrm{GeV}}\right)^{7/6}.$$ (9.13)

If PQ symmetry breaking happens before inflation, each region of space with independent θ_i rapidly expands, such that even in the present day, the regions are not in causal contact. Therefore, the entire visible universe today originated from a single patch of space corresponding to the same θ_i value in the entire universe. Given the latest measurement of the cold dark matter (CDM) density of $\Omega_{\mathrm{CDM}} = 0.12$ [22], an axion mass on the order of 1 μeV would account for 100% of the dark matter in the universe.

If PQ symmetry breaking happens after inflation, then different regions of space would take on different values of θ_i. The total relic density would be the average density of each patch of space. Assuming a uniform distribution of θ_i, the variance of the angle is $\langle\theta_i^2\rangle = \pi^2/3$. Given the relic energy density today, assuming axion makes up all of the dark matter density, would result in a PQ symmetry breaking scale of $f_a \sim 10^{12}\,\mathrm{GeV}$, corresponding to an axion mass of 10^{-6}–10^{-5} eV. For this reason, many axion experiments have been searching for axions in this mass range for the past decade. However, as will detailed in the next section, the null result and slow mass scan rate of current experiments have motivated the BREAD experiment in this book to search for heavier axions, favored by several theoretical scenarios [316, 324, 325], with a broadband detection technique.

9.2.5 Axion Detection

The most commonly explored interaction is the axion-photon interaction:

$$L_{a\gamma\gamma} = -\frac{1}{4}g_{a\gamma\gamma}F_{\mu\nu}\tilde{F}^{\mu\nu}a$$
$$= g_{a\gamma\gamma}a\mathbf{E}\cdot\mathbf{B}$$ (9.14)

where $g_{a\gamma\gamma}$ is the axion-photon coupling given by Equation 9.9 for QCD axions. For axion-like particles, $g_{a\gamma\gamma}$ is a free parameter independent of mass. $F_{\mu\nu}$ is the EM field strength tensor, $\mathbf{E}$ is the electric field, and $\mathbf{B}$ the magnetic field.

With this interaction included, the Maxwell's equations are modified as follows [326],

$$\nabla\cdot\mathbf{E} = \rho - g_{a\gamma\gamma}\mathbf{B}\cdot\nabla a$$
$$\nabla\cdot\mathbf{B} = 0$$
$$\nabla\times\mathbf{E} = -\frac{\partial\mathbf{B}}{\partial t}$$
$$\nabla\times\mathbf{B} = \frac{\partial\mathbf{E}}{\partial t} + \mathbf{J} - g_{a\gamma\gamma}\left(\mathbf{E}\times\nabla a - \frac{\partial a}{\partial t}\mathbf{B}\right). \tag{9.15}$$

The gradient of the axion field (∇a) is small, because it arises from the axion velocity dispersion. However, axions are produced at rest, so it only has a small velocity dispersion primarily from a history of gravitational interactions. Therefore, the strongest signal to search for axion is the effective axion source current in the final term of Ampère's Law: $\mathbf{J} = g_{a\gamma\gamma}\frac{\partial a}{\partial t}\mathbf{B}$.

Using the expression of the axion field when its stably oscillating, given by Equation 9.12, the effective source current can be written as:

$$\mathbf{J_a} = g_{a\gamma\gamma}\sqrt{2\rho_{DM}}\cos(m_a t)\mathbf{B} \tag{9.16}$$

where ρ_{DM} is the local dark matter density, m_a is the axion mass, and $\mathbf{B}$ is the external magnetic field. As shown in the equation, the frequency and strength of the oscillating effective current depends on the mass and coupling respectively, but the existence of this effect is independent of the axion model. Since the possible axion mass spans orders of magnitude, different detection technology are used to enhance and detect the small EM field induced by different axion masses.

This interaction of the axion with photons in an external magnetic field is called the Primakoff effect, as shown in Figure 9.4 (left). This process is analogous to the kinetic mixing of dark photons (A') with SM photons [308, 309], as shown in Figure 9.4 (right), except the dark photon-photon conversion persists even without an external magnetic field. Analogously, the kinetic mixing between the dark photon and the SM photons is given by the interaction Lagrangian:

$$L_{A'} = -\frac{1}{4}\kappa F'_{\mu\nu}F^{\mu\nu}a \tag{9.17}$$

where κ is the A'-SM kinetic mixing parameter, $F^{(\prime)}_{\mu\nu}$ is the SM (dark) photon field strength. The A'-SM kinetic mixing also contributes to the effective source current

Figure 9.4: Diagram of the conversion of an axion to a SM photon under an external B field (left) and the conversion of a dark photon to a SM photon (right).

in the Ampère's Law:

$$\mathbf{J_{A'}} = \kappa m_{A'} \sqrt{2\rho_{DM}} \cos(m_{A'}t)\hat{\mathbf{n}} \tag{9.18}$$

where $\hat{\mathbf{n}}$ is the polarization of the dark photons, and $m_{A'}$ is the mass of the dark photon.

The most sensitive detection strategy today is the radio-frequency resonant-cavity haloscope [326–329], where ADMX [330–336], CAPP [337–339], HAYSTAC [340–342] probe QCD axions with masses between 1.8–2.4 μeV. However, this strategy has a long-standing disadvantages from narrow-band tuning to the unknown axion mass, and impractical high-mass scaling for $m_a \gtrsim 40\mu$eV. Scan rates fall steeply with the photon frequency $R_{\mathrm{scan}} \sim \nu^{-\frac{14}{3}}$ [343] and the number of required resonators scales unfavorably with the effective volume $\sim m_{\mathrm{DM}}^3$. Proposed dielectric haloscopes could probe higher masses from 40–400 μeV [344–346] and topological insulators can probe 0.7–3.5 meV. However, significant gaps in sensitivity to axions still persist across 0.1 meV–1 eV, favored by several theoretical scenarios [316, 324, 325], motivating new broadband detection techniques that can probe heavier axions.

The Broadband Reflector Experiment for Axion Detection (BREAD) proposes a novel experimental design to search for multiple decades of DM masses from $\sim \mu$eV to eV without tuning. The experiment uses a novel cylindrical dish resonator that fits in a solenoidal magnets. Axions or dark photons are converted to the photons at the metallic surface, regardless of their masses. These photons are then focused by a parabolic focusing reflector onto a low noise single photon counting detector. One of the promising single photon counting detectors that will be used for the BREAD experiment is SNSPDs, that are sensitive to 0.04 to 1 eV axions and dark photons, due to their sensitivity to 1–30 μm photons. A first stage pilot dark photon experiment using SNSPD is being planned for at Fermilab and ongoing work is being carried out to characterize the SNSPDs to commission the pilot experiment.

9.3 Novel Parabolic Reflector for Broadband Axion Detection

The section details the conceptual design and preliminary sensitivity estimate of the BREAD experiment, that can search for multiple decades of axion or dark photon mass without tuning.

As mentioned in the previous section, the axions (in the presence of an external magnetic field) or dark photons could induce a nonzero source current, given by Equation 9.16 and 9.18. The nonzero source current would induce a small EM field that could cause a discontinuity at the interface of media with different electric permittivity, such as a conducting surface in vacuum. To satisfy the boundary condition that $E_\parallel = 0$, a compensating EM wave with amplitude $|\mathbf{E}|$ and frequency m_{DM} must be emitted perpendicular to the surface. The EM waves transmit power $P = \frac{1}{2}|\mathbf{E}|^2 A_{\mathrm{dish}}$ for a dish area of A_{dish}. The emitted power for axion (a) and dark photons (A') are:

$$P_a = \frac{1}{2}\rho_{DM}(B_{\mathrm{ext}}^{\parallel} g_{a\gamma\gamma}/m_a)^2 A_{\mathrm{dish}}$$

$$P_{A'} = \frac{1}{2}\rho_{DM}\kappa^2 A_{\mathrm{dish}}\alpha_{\mathrm{pol}}^2 \tag{9.19}$$

where the factor $\alpha_{\mathrm{pol}}^2 = \sqrt{2/3}$ averages over A' polarizations [309]. In the photon counting regime, the rate of emitted photon is given by $R_{\mathrm{DM}} = P_{\mathrm{DM}}/m_{\mathrm{DM}}$, since the emitted photon energy equals the DM mass. The rate of photons for axion and dark photons are:

$$R_a = \frac{1}{2}\rho_{DM}(B_{\mathrm{ext}}^{\parallel} g_{a\gamma\gamma})^2 A_{\mathrm{dish}}/m_a^{-3}$$

$$R_{A'} = \frac{1}{2}\rho_{DM}\kappa^2 A_{\mathrm{dish}}\alpha_{\mathrm{pol}}^2/m_{A'}. \tag{9.20}$$

BREAD proposes a cylindrical barrel as the emitting surface and a novel reflector geometry comprising a coaxial parabolic surface of rotation around its tangent. This focuses the emitted photon to a photosensor located on-axis at the parabola's vertex as shown in Figure 9.5. In principle the DM-photon conversion can also occur on the parabolic surface, but these photons are not focused to the vertex. For a barrel with radius R and length $L = 2\sqrt{2}R$, the emitting area is $A_{\mathrm{dish}} = 2\pi RL = 4\sqrt{2}\pi R^2$. This cylindrical geometry and aspect ratio suit the enclosure in conventional high-field solenoid magnets and ensure that the magnetic field is parallel to the emitting surface.

The DM-photon conversion occurs independent of the DM mass, in principle allowing searches across several mass decades in single runs. However, practically,

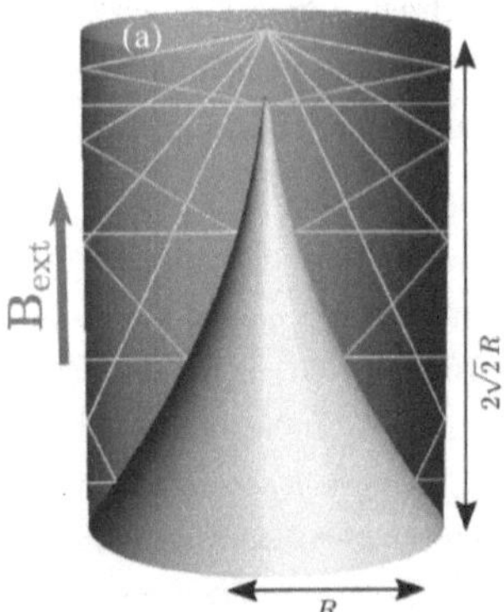

Figure 9.5: BREAD reflector geometry: rays (yellow lines) emitted perpendicular from the cylindrical barrel, which is parallel to an external magnetic field B_{ext} from a surrounding solenoid (not shown) and focused at the vertex by a parabolic surface of revolution, reprinted from [347].

the sensitivity also depends on the focusing of the reflector and the photosensor sensitivity. At high masses, the sensitivity is limited by the focusing effect, which broadens the focal spot and reduces the geometric acceptance of the photons due to the finite photosensor size. In the high-mass limit $\lambda_{dB} \ll R$, DM-photon conversion occurs incoherently. The DM-halo velocity $v \simeq 10^{-3}$ smears out the focal spot size [348–350], since the photons are emitted at an angle proportional to the DM-halo velocity with respect to the perpendicular direction. An illustration showing the focusing effect with 100 times exaggerated effect is shown in Figure 9.6 (left). The detector acceptance for different focal spot sizes is shown in Figure 9.6 (right) for different DM wind direction. For a pilot experiment with $R = 0.2$ m, a 1×1 mm^2 photodetector would have a detector acceptance of 45% (75%) assuming z (x/y) DM-wind alignment.

To collect the photons that are focused to a vertex, a photon detector with broad spectral response, low noise, and $\sim$ mm^2 size active area is needed. Due to the large active area of 1 mm^2 [264], low dark count rate of 6×10^{-6} cps [255], and unique and broad spectral response from 0.04 to 1 eV [282], SNSPD satisfies all of the photosensor requirements for BREAD. Therefore, the first phase of the BREAD experiment, which is a pilot dark photon experiment that does not require a magnetic field yet, will use SNSPD as the photon detector. The possibility and sensitivity

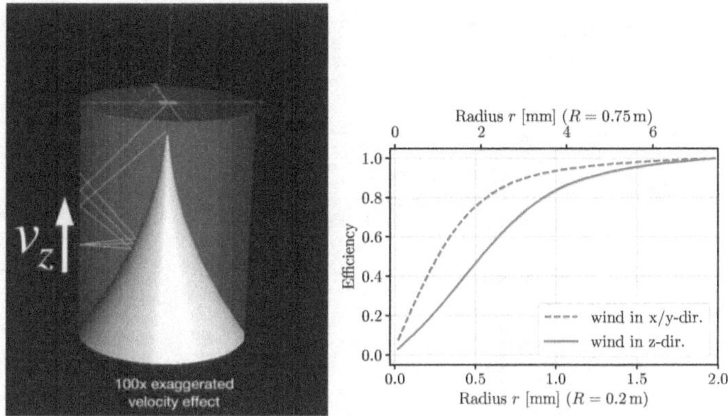

Figure 9.6: An illustration showing the focusing effect with 100 times exaggerated effect (left). The detector acceptance with respect to focal spot sizes for different DM wind direction (right), reprinted from [347]. The top and bottom x-axes correspond to different barrel radii (R) for the pilot and full-scale experiment, respectively.

of the BREAD experiment using other photon detectors such as kinetic inductance detectors, transition edge sensor, and quantum capacitance detectors, are discussed in Ref. [347]. Each detector has their own advantages and disadvantages, in terms of the spectral response, dark count rate, and active area, but SNSPD is uniquely positioned to probe the DM mass range of 0.04–1 eV and has a large active area that already satisfies the requirement for the pilot BREAD experiment.

The pilot experiment will have an $R = 0.2$ m and $A_{\text{dish}} = 0.7\,\text{m}^2$, with the barrel constructed from aluminum. A schematic diagram of the proposed experimental design for the pilot experiment is shown in Figure 9.7. The cylindrical conducting surface and the parabolic reflector will be cooled to 4 K to suppress thermal noise and the SNSPD will be cooled to a lower temperature at sub-Kelvin level. The photon sensor will be installed on a piezoelectric motion stage to fine-tune the sensor position at the focal point and to move the sensor to an off-focus position for an in-situ noise measurement. Additionally, a monochromatic laser or bandpass-filtered blackbody source can inject photons via a small hole in the barrel for in-situ calibration and monitoring of the reflector-photosensor setup. The pilot dark photon search will have a runtime of about 10 days, which will serve as a proof-of-principle for the BREAD experiment and already provide sensitivity to previously unconstrained parameter space, as discussed in the next paragraph. In the longer term, a full-scale experiment with $R = 0.75$ m and $A_{\text{dish}} = 10$ m^2, will be hosted inside a 10 T magnet.

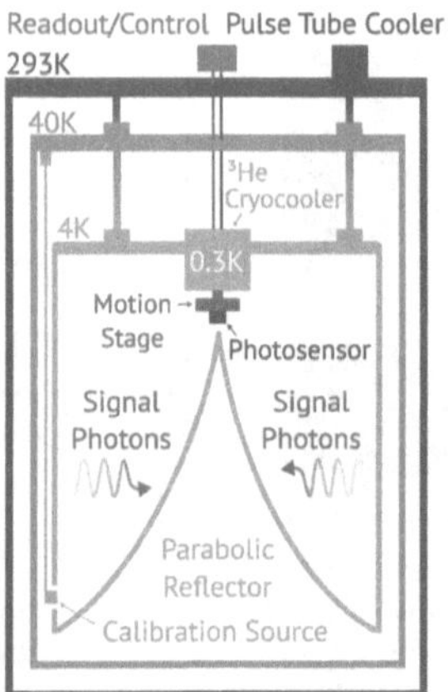

Figure 9.7: The schematic setup in cryostat for the pilot dark photon experiment is shown, reprinted from [347].

Give the DM-induced rate of emitted photons (R_{DM}) from Equation 9.20, the significance Z can be used as a figure-of-merit to estimate the sensitivity of the BREAD experiment. The significance Z given a runtime Δt, for a photosensor with dark count rate DCR and detection efficiency ϵ is estimated as [347]:

$$Z = \frac{N_{\text{signal}}}{\sqrt{N_{\text{noise}}}} = \frac{\epsilon R_{\text{DM}} \Delta t}{\sqrt{\text{DCR} \Delta t}}. \tag{9.21}$$

Requiring $Z = 5$ for DM reach implies the coupling sensitivity is [347]:

$$\left\{ \begin{array}{c} \left(\dfrac{g_{a\gamma\gamma}}{10^{-12}}\right)^2 \\[2mm] \left(\dfrac{\kappa}{10^{-15}}\right)^2 \end{array} \right\} = \left\{ \begin{array}{c} \dfrac{3.0}{\text{GeV}^2}\left(\dfrac{m_a}{\text{meV}}\right)^3\left(\dfrac{10\,\text{T}}{B_{\text{ext}}}\right)^2 \\[2mm] 11.9\dfrac{2/3}{\alpha_{\text{pol}}^2}\dfrac{m_{A'}}{\text{meV}} \end{array} \right\}\left(\dfrac{\text{hour}}{\Delta t}\right)^{1/2}$$
$$\times \frac{10\,\text{m}^2}{A_{\text{dish}}}\frac{Z}{5}\frac{0.5}{\epsilon}\left(\frac{\text{DCR}}{10^{-2}\,\text{Hz}}\right)^{1/2}\frac{0.45\,\text{GeV/cm}^3}{\rho_{\text{DM}}}. \tag{9.22}$$

The factor $\alpha_{\text{pol}} = \sqrt{2/3}$ averages over A' polarizations [309] and $\rho_{\text{DM}} = 0.45$ GeV$/$cm^3 is assumed for the local dark matter density.

Using Equation 9.22, we project the BREAD sensitivity to dark photon, as shown in Figure 9.8 (left). The projection assumes that SNSPD has a DCR of 10^{-4} cps, a flat 50% detection efficiency over the entire spectral range between 0.04–1 eV, and all photons are focused onto the detector. Existing constraints following Ref. [351] include stellar astrophysics [352, 353], cosmology [310, 354, 355], and $\gamma \rightarrow A'$ conversion that includes laboratory probes [356, 357]. With just 1 day runtime of the pilot dark photon experiment with $A_{\mathrm{dish}} = 0.7\,\mathrm{m}^2$, the solid red line shows the BREAD experiment can surpass existing constraints on the kinetic mixing by at least an order of magnitude for 0.04–1 eV masses. More importantly, the pilot experiment probes higher masses than existing haloscope experiments such as ADMX [331, 332], CAPP [337–339], HAYSTAC [342, 346], transmon qubit [358], and WIS-PDMX [359], whose results are recasted for dark photon following Ref. [351].

Similarly, the axion sensitivity is calculated using Equation 9.22, as shown in Figure 9.8 (right). Existing constraints [351] additionally include results from the CAST helioscope [360, 361], telescopes [362, 363], neutron stars [364–366], alongside ORGAN [367], QUAX [368, 369], RADES [370], and URF [327, 329, 371] haloscopes. Running the full-scale experiment with 10 days already allow us to surpass existing limits from CAST in certain mass ranges, but to practically probe the QCD axion models requires longer runtime and lower sensor noise. As shown in the figure, with 1000 days of runtime we can reach the KSVZ model [296, 372] in mass ranges 0.04–0.1 eV, but to reach KSVZ and DFSZ model [299] in the full mass range 0.04–1 eV requires 1000 days of runtime with a DCR of 10^{-6} cps. Reaching the challenging DCR of 10^{-6} cps for large area SNSPDs requires significant sensor development in the next years.

9.4 Characterization of SNSPD for the Pilot Dark Photon Experiment

We are currently planning for a pilot dark photon experiment with a $0.7\mathrm{m}^2$ barrel using SNSPD as photon detector at Fermilab that would surpass existing dark photon coupling constraints by over a decade with one-day runtime. As shown in Section 9.3, to successfully commission the experiment, it is important to fully understand the signal efficiency in different photon energy range and the dark count rate of the SNSPDs. In this section, my work on characterizing the SNSPD is discussed in detail.

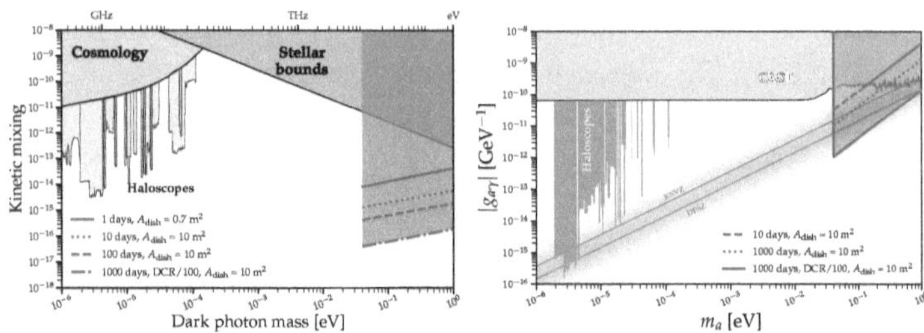

Figure 9.8: The projected BREAD sensitivity for axion (left) and dark photon (right). This assumes a DCR of 10^{-4} cps, a flat 50% efficiency over the entire spectral range, dish area of 10m^2, and all photons are focused onto the detector. The limits are calculated when $Z = 5$. Blue shading shows existing constraints from Ref. [351]. Benchmark axion predictions include QCD axion models KSVZ [296, 372] and DFSZ [299]. The limits are produced with code adapted from this package [373].

9.4.1 8-channel mm^2 SNSPD

The pilot dark photon experiment will use an 8-channel mm^2 SNSPD fabricated at the Jet Propulsion Lab, corresponding to film 2 in Ref. [264]. As discussed in Section 9.3, the 1 mm^2 active area can already capture 45-75% of the signal, depending on the exact direction of the dark matter wind.

The 8-channel SNSPD was designed as 1 μm-wide meanders with a 1 mm^2 active area and a fill factor of 0.25. The area was chosen to be as large as possible while still fitting into a single write field of the electron-beam lithography tool, to avoid stitching errors. To investigate uniformity and yield, the detector was divided into eight pixels, each with an active area of 1×0.125 mm. Each pixel is connected to ground on one side and to an individual readout line on the other side.

The 8-channel SNSPD is made from WSi films and deposited onto an oxidized Si(100) substrate with a 240 nm-thick thermal oxide. It was sputtered from a W$_{50}$Si$_{50}$ target in an argon atmosphere at a pressure of 5 mTorr and a sputter power of 130 W. After fabrication, the composition of the SNSPD was analyzed using Rutherford Backscattering Spectrometry and determined to be 70% tungsten and 30% silicon. The sheet resistance is 570 Ohm, measured using a four probe setup at room temperature. The critical temperature is 3.25 K, determined by the temperature at the inflection point of the measured resistance-temperature curve. A picture of the SNSPD used in the rest of this chapter is shown in Figure 9.9.

Figure 9.9: A picture of the SNSPD attached to the cold finger of the cryostat.

9.4.2 Experimental Setup

9.4.2.1 Cryogenic System

To operate the SNSPDs at superconducting temperature of a few Kelvin, we use an adiabatic demagnetization refrigerator (ADR) that has a base temperature of 50 mK and has windows throughout the different temperature stages to allow for external illumination. The compact Rainer Model 103 ADR designed by High Precision Devices, Inc. is being used to operate and characterize the SNSPDs. It provides a relatively large experimental space of 26 cm in diameter and 25 cm in height. The cryostat consists of four temperature stages: 40 K, 4 K, 1 K, and 50 mK. The cryostat frame, as shown in Figure 9.10, consists of three stage plates (room temperature, 40 K, and 4K) connected by thermally isolating supports. Below the 4 K plate, lies the 1 K stage and the cold finger at 50 mK. Each plate includes a series of pass-throughs for experimental and thermometer wiring. Additionally, radiation shields, which are not shown in the figure, are connected to the 40 K and 4 K plates to block blackbody radiation from reaching the low temperature experimental volume. The radiation shields can be configured to have opened windows to allow external illumination to reach the tested device, as detailed in Section 9.4.4.

The cooling of the 40 K and 4 K is provided by a Cryomech PT407 two-stage pulse-tube refrigerator (PTR), which is closed-loop system that provides cooling without using any liquid cryogens. Helium is used as the working gas, due to its monotonic ideal gas properties and low condensation temperature of 4.2 K. The coldest stage

Figure 9.10: The opened ADR with SNSPD attached is shown. The cryostat frame consists of three stage plates (room temperature, 40 K, and 4K) connected by thermally isolating supports. Below the 4 K plate, lies the 1 K stage to the right of the picture and the cold finger at 50 mK, where the SNSPD is attached. Each plate includes a series of pass-throughs for experimental and thermometer wiring.

temperatures (1 K and 50 mK) are provided by the ADR in the cryostats. The two-stage ADR contains a superconducting 4 T magnet, a gadolinium gallium garnet (GGG) paramagnetic salt pill, a ferric ammonium alum (FAA) paramagnetic salt pill, a Hiperco 50 magnetic shield, and a Kevlar suspension system. The ADR generates cooling through adiabatic demagnetization of the paramagnetic salt pills, which is thermally isolated from warmer stages through the Kevlar suspension system. Cooling is achieved by transferring the entropy between the paramagnetic materials thermal and magnetic entropy components in two steps: isothermal magnetization and adiabatic cooling. In the first step of isothermal magnetization, a magnetic field is applied to the salt pills, where the spins of the material align, decreasing the magnetic entropy. During this step the ADR is attached to the PTR at about 4 K, thus this step is isothermal and the overall entropy of the system decreases. The adiabatic cooling process occurs after thermal equilibrium is complete (1-2 hours) and the salt pill is disconnected from the PTR through a heat switch. At the same time the magnetic field is reduced, so the spins of the paramagnetic material will slowly randomize in directions, resulting in an increase in magnetic entropy and a decrease in thermal entropy and temperature towards the ordering temperature of the material. The two types of salt materials that are used in the ADR provide cooling for two separate stages. FAA has an ordering temperature near 26 mK [374] and provides cooling to the 50 mK stage, while GGG has an ordering temperature near 0.38 K [374] and provides cooling to the 1 K stage. ADRs offer only discontinuous cooling, which means that once the magnetic entropy of the salt pill reaches its maximum, the desired temperature can no longer be maintained. At this point, the system needs to be recycled. The hold time of the ADR with the heat load from the mm^2 SNSPD is typically 12-24 hours, for an operating temperature of 0.2–0.8 K. The ADR is usually operated at high vacuum state of 10^{-7} mbar to ensure the absence of water molecules that could freeze in the cryostat at low temperature and the absence of heat transfer between different temperature stages through gas molecules.

9.4.2.2 Readout and Data Acquisition

The 8-channels in the SNSPD are connected to individual readout lines. Therefore, each channel can be connected through RF cables to room temperature. Typical readout of SNSPDs is based on room temperature, low-noise RF amplifiers with AC-coupling to prevent SNSPDs from reaching the critical current due to an early onset of latching. However, if the recovery timescale of the SNSPD is significantly longer than the timescale corresponding to the low-frequency cut-off of the amplifier, there is a significant undershoot in the readout signal, resulting in additional current being sent to the detector during the recovery phase, causing the detector to latch. The μm-wide device that we are using has a sheet-inductance of 650 pH/square and $L_K = 17.5$ μH for the device, thus the reset timescale is $O(1\ \mu s)$, much longer than typical radio-frequency amplifiers that cut off at $O(100\text{ kHz})$, corresponding to $O(100\text{ ns})$. To overcome this limitation, a DC-coupled amplifier operating at 40 K is being used, as shown in Figure 9.11. The DC-coupled amplifier board also uniquely integrates signal readout and SNSPD biasing to a single board.

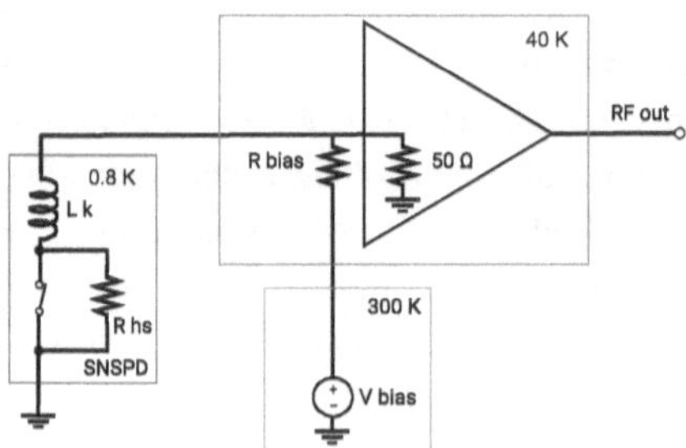

Figure 9.11: The readout and biasing scheme used, reprinted from [264]. The cryogenic amplifier board is represented in the 40 K box, consisting of a resistive bias-tee and a two-stage cryogenic amplifier. The SNSPD is represented as a variable resistor R_{hs} in series with an inductor with kinetic inductance L_k. R_{bias} is 10 kΩ in the setup.

However, DC-coupled amplifiers can have DC offsets in the bias current caused by series resistance. The DC offsets could also change slightly between each cool down. Therefore, in each cool down, we measure the current-voltage characteristic (IV-curve) to calculate the DC offset to correct for the bias current for the SNSPD before performing any other measurements. The IV-curve can be obtained by measuring the voltage across of the SNSPD (V_{sense}) as the bias voltage (V_{bias}) varies, as shown in Figure 9.12 (left). R_{hs} is then extracted from a linear fit in the superconducting regime with the following relationship:

$$V_{\text{bias}} = V_{\text{sense}}[1 + 10 \text{ k}\Omega \times (\frac{1}{50\Omega} + \frac{1}{R_{\text{hs}}})]$$

(9.23)

where the 10 kΩ corresponds to R_{bias} in the setup. Subsequently, the bias current I_{bias} that passes through R_{bias} can be determined:

$$I_{\text{bias}} = \frac{V_{\text{bias}}}{10k\Omega + \frac{50\Omega \times R_{\text{hs}}}{50\Omega + R_{\text{hs}}}}.$$

(9.24)

Plotting I_{bias} with respect to V_{sense} then allow us to measure the DC offset (I_{offset}) in the bias current, determined by the mean of the two bias currents (positive and negative) where the transition between superconducting and normal state occurs, as shown in Figure 9.12 (right). If the transitions occur at the same absolute bias current on the positive and negative end, then there is no DC offset. The corrected bias current through the SNSPD is then written as:

$$I_{\text{SNSPD}} = (I_{\text{bias}} + I_{\text{offset}}) \times \frac{50\Omega}{50\Omega + R_{\text{hs}}}.$$

(9.25)

The bias currents that are calculated and shown in the rest of this chapter correspond to the corrected bias current (I_{SNSPD}).

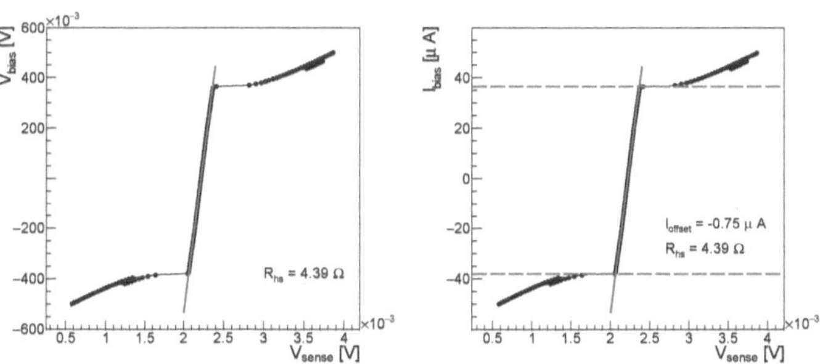

Figure 9.12: The bias voltage (V_{bias}) with respect to sense voltage (V_{sense}) is shown on the left. The bias current (I_{bias}) with respect to sense voltage is shown on the right.

The readout and biasing scheme, including room temperature infrastructure is also illustrated in Figure 9.13. As shown in Figure 9.13, the SNSPD signal from each channel is readout through a single cable from the SNSPD, through the 1 K and 4 K stages through feed-throughs, to the input of the amplifier board attached at the 40 K stage. The SNSPD signal is then amplified and readout to a digitizer or scope at room temperature. The SNSPD bias current is also provided through the same cable between the SNSPD and amplifier board. The SNSPD bias current for each channel along with the amplifier bias current are provided through a biasing breakout board at room temperature, which can take inputs from a number of voltage sources to provide bias current for up to four SNSPD channels. A picture of the biasing breakout board is shown in Figure 9.14.

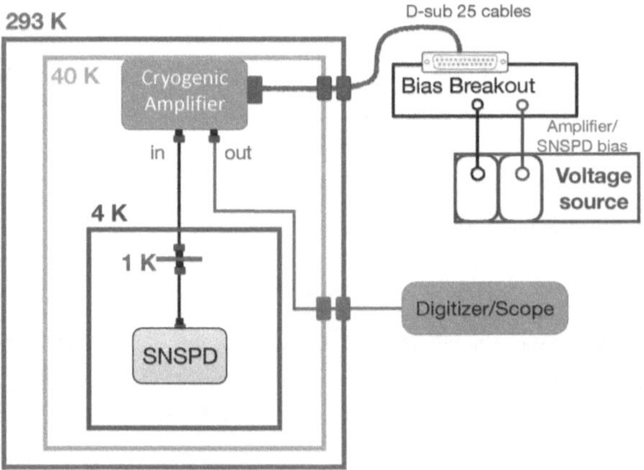

Figure 9.13: A simplified schematic diagram of the cryogenic and electronic setup for a single channel is shown. The SNSPD signal is readout through a single cable from the SNSPD to the input of the cryogenic amplifier, which is then amplified and readout to room temperature. The SNSPD bias current is also provided through the same cable between the SNSPD and amplifier board. The SNSPD and amplifier bias current are provided through the biasing breakout board at room temperature. All of the bias currents are grouped into D-sub cables, consisting of 25 electrical connections, to be transmitted from the biasing breakout board at room temperature to the cryogenic amplifier.

The cryogenic amplifier is a 4-channel amplifier board with two amplification stages. A picture of the amplifier is shown in Figure 9.15 (left). The first stage is based on a low noise high-electron-mobility transistor fabricated by Skyworks [375] and the second stage is based on silicongermanium amplifier manufactured by Qorvo. The

Figure 9.14: The room-temperature biasing breakout board is shown.

total gain has been measured to be 30 dB from 10-500 MHz using a vector network analyzer, as shown in Figure 9.15 (right).

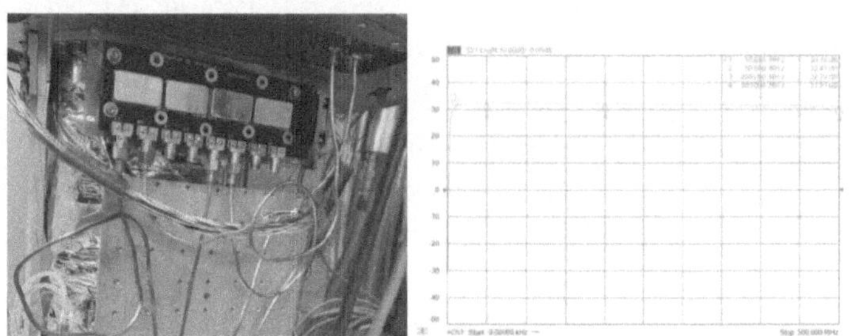

Figure 9.15: The 40 K cryogenic 4-channel amplifier board (left) and the characterization of its gain with respect to frequency (right) are shown.

The readout cables that transmit the SNSPD signals need to be able to operate in cryogenic temperature. These cryogenic cables need to have high electrical conductivity to transmit the signal efficiently, low thermal conductivity and heat capacity to preserve the ultra-cold temperature, and low thermal expansion coefficient to remain mechanically stable. Therefore, the cables between the SNSPD and 1 K stage need to be special cryogenic cables that are made of niobium-titanium. The other cables at higher temperatures have less stringent requirements. At the time of writing, we are limited by the number of available niobium-titanium cables, the characterization presented in the rest of this section corresponds to only one channel (channel 3) in the 1mm^2 SNSPD. There is ongoing effort to improve the readout to be able to readout all 8 channels simultaneously.

At room temperature, a Swabian Time Tagger [376], a multi-channel time-to-digital converter that can tag up to 70M tags/s, is used to measure the dark count rate and photon count rate based on a predefined threshold. The threshold is defined to be 80 mV, based on the analog shape of the SNSPD signal pulse observed in an oscilloscope.

9.4.3 Dark Count Rate Measurement

The dark count rate of the SNSPD is one of the most important characteristics to be measured, which determines the background level of a DM experiment. The dark count rate is measured by counting the number of SNSPD pulses per second using the Swabian Time Tagger. To be able to measure the lowest dark count rate possible, we make sure that the radiation shield at each temperature is light tight, to prevent black-body radiation at high temperature. As shown in Figure 9.16, light-tight radiation shields are added at 40 K and at room temperature, and additionally a detector lid is added to the SNSPD to prevent background thermal photons from reaching the SNSPDs.

Figure 9.16: The radiation shield (left) at 40 K and the detector lid (right) are shown.

The radiation shields can lower the dark count rate by five orders of magnitude, allowing us to achieve an unprecedentedly low dark count rate of 10^{-3} cps at 0.2 K, measured for the first time on micron-wide, large-area SNSPDs. The dark count rate with respect to the bias current is shown in Figure 9.17. The measured DCR has two distinct regions. The exponential component at higher bias current corresponds to intrinsic dark count rate from internal thermal fluctuations. The origin of the intrinsic dark counts is expected to be current-assisted unbinding of vortex-antivortex pairs [377], which has an exponential dependence on the bias current. The observed exponential behavior of intrinsic dark counts, is also in agreement with previous studies for narrower wires [377]. The flat component at lower bias current approaching 10^{-3} cps suggests that residual background photons are reaching the SNSPD. The residual background photons may be further mitigated with a more light-tight detector enclosure that is underdevelopment to reject photons that might be guided down the coaxial cables connected to the sample box.

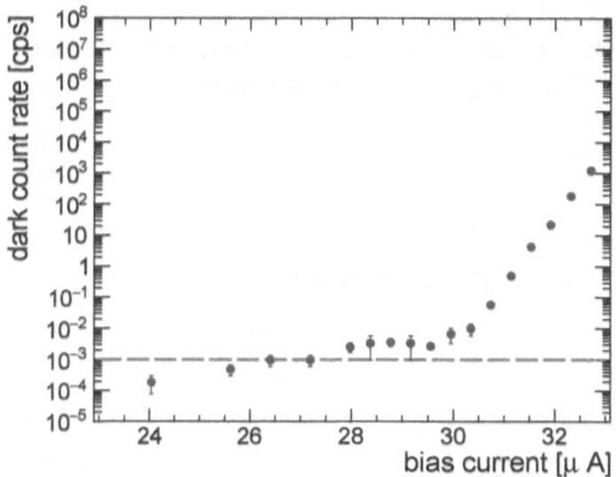

Figure 9.17: The dark count rate with respect to the SNSPD bias current is shown.

ADR allow us to operate the SNSPDs at lower temperature and understand for the first time the temperature dependence of the dark count rate on large-area SNSPDs. Figure 9.18 shows the DCR with respect to the bias current, measured at different operating temperature. As shown in the figure, as the operating temperature decreases, the DCR shifts significantly to the right. Therefore, given the same bias current, operating at lower temperature significantly decreases the dark count rate, demonstrating the advantage to operate at as low temperature as possible for DM experiments. This temperature-dependence that we have observed is in agreement with previous studies for narrower wires [377]. The DCR also decreases if we operate at a lower bias current, but generally the detection efficiency or the photon count rate also decreases at lower bias current, so it is not desirable to tune the bias current significantly, as discussed in Section 9.4.4.

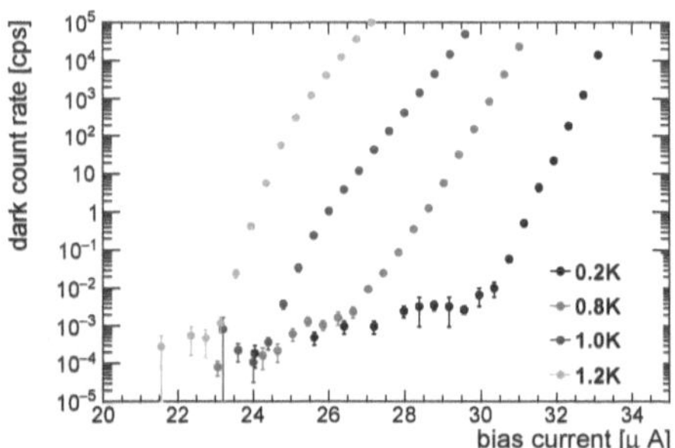

Figure 9.18: DCRs measured at different operating temperature is shown. Given the same bias current, operating at lower temperature significantly decreases dark count rate. This demonstrates the advantage to operate at as low temperature as possible for DM experiments.

9.4.4   Photon Count Rate Measurement

In this section, the effort towards measuring the signal efficiency and the measurement of the normalized photon count rate (PCR) with external laser excitation is discussed. To allow a photon from an external laser at room temperature to reach the SNSPD, all windows that were shielded at different temperature now need to be opened with optical filters added to keep the background photon rate low. Fiber-based laser diodes placed at room temperature are coupled to the cryostat through a reflective collimator (RC08SMA [378]) that has high reflectivity within the 450 nm–20 μm wavelength range and outputs a collimated beam of 8.6 mm in diameter. The PCR has been studied with four fiber-based laser diodes [379] from QPhotonics with wavelengths of 635, 1060, 1300, and 1650 nm to understand the SNSPD response to different photon wavelengths. A schematic diagram of the setup is shown in Figure 9.19 and a picture of the setup from outside of the cryostat is shown in Figure 9.20.

To be able to accurately measure the photon count rate and reduce the background count rate from thermal photons, a few optical filters are added at different temperature stages. A custom cryogenic short-pass filter [380] composed of a N-BK7 glass window with optical coating deposited by Andover Corp., is added to the 40 K stage to filter out photons with wavelengths between 1.9 and 4.5 μm with an optical density of 3. The filter rejects wavelengths shorter than 3 μm through the reflective optical coating, and attenuates longer wavelengths through material absorption in

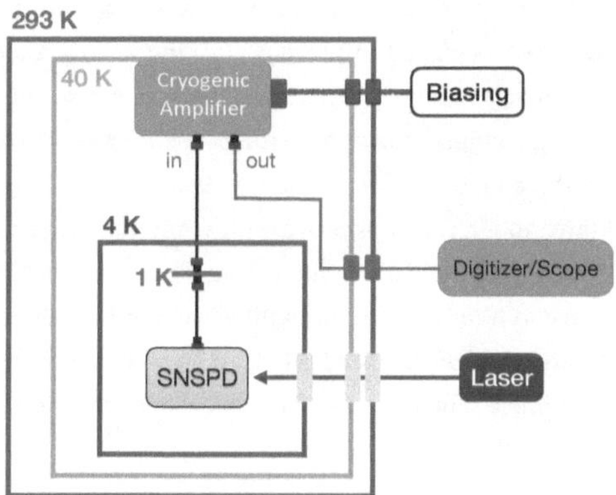

Figure 9.19: A schematic diagram of the setup to measure PCR is shown. Room temperature fiber-based laser diode is coupled to the 293 K stage through a reflective collimator. The output collimated beam travels through the optical filters at different temperatures stages that are shown in gray to finally reach the SNSPD.

Figure 9.20: A picture of the laser diode controller, fiber-based laser diode, and reflective collimator attached on the ADR is shown.

the 12.7-mm-thick N-BK7 glass substrate. Additionally, since this is a custom filter, a custom clamp was also designed to prevent cracking at low temperature due to the difference in thermal contraction between glass and metal, while maintaining good contact. A spring clamped mirror mounting scheme was adapted, as shown in Figure 9.21, where three G10 spacers provide vertical force to keep the glass window in place and three mounting clips provide only minimal force to keep the glass window from falling. A neutral density filter from Thorlabs (NENIR40A [381]) is added at the 4 K stage to filter out photons with wavelengths between 1 and 2.6 μm with an optical density of 4. Background rate is reduced to a few counts per second after the filters are applied. The pictures of the two filters attached on the ADR are shown in Figure 9.22.

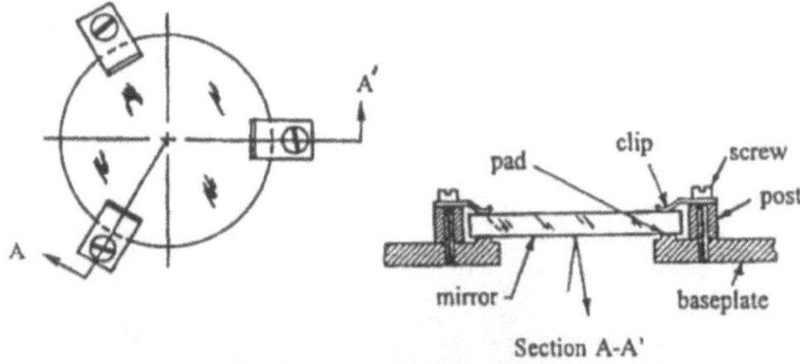

Figure 9.21: A simple spring clamped mirror mounting. (Reprinted from [382])

The normalized photon count rate for 635 nm photon overlayed with the DCR, both measured at 0.2 K, are shown in Figure 9.23. The normalized photon count rate is calculated by the photon count rate subtracted by the background count rate, measured in the same setup by turning the laser off, and then normalize the plateau to 1. The shape of the normalized photon count rate is expected to be the same as that of the detection efficiency curve, except the normalization is not known. The plateau at high bias current demonstrates that the detector internal detection efficiency is saturated and the bias current range 26–29 μA is optimal for operation with a large efficiency and a low dark count rate.

The normalized photon count rate measured at 0.2 K for different wavelengths of 635, 1060, 1300, and 1650 nm are shown in Figure 9.24 (left). As shown in the plot, the internal detection efficiency is saturated for 635 and 1060 nm and the

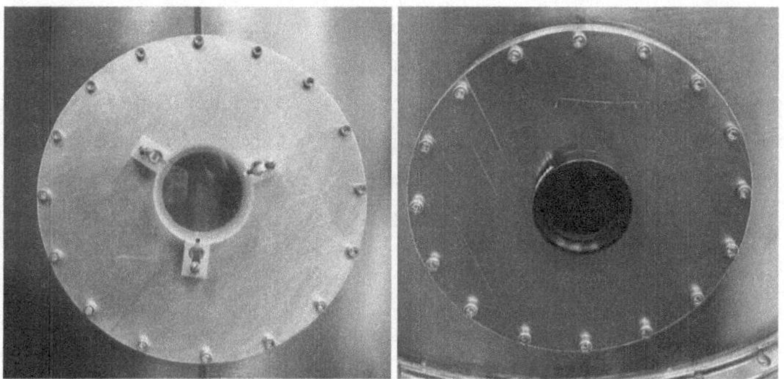

Figure 9.22: The custom short-pass filter (left) and neutral density filter (right) added at the 40 K and 4 K stage, respectively, to filter out background thermal photons to be able to accurately measure the PCR.

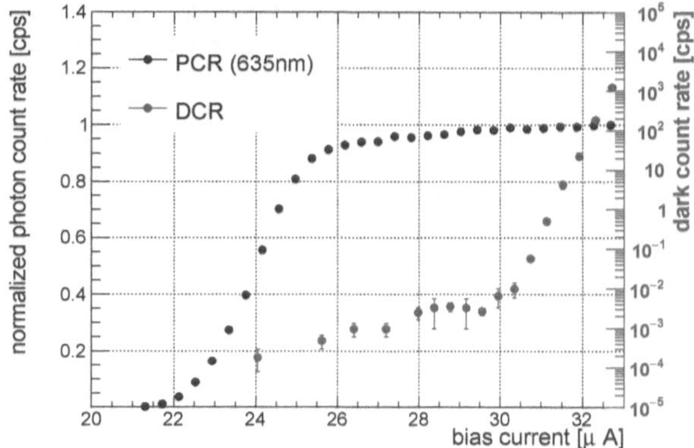

Figure 9.23: The normalized photon count rate for 635 nm photon overlayed with the DCR, both measured at 0.2 K.

saturation occurs at a lower bias current for higher photon energy. It has been recently demonstrated with nm-wide SNSPDs that lower photon energy sensitivity can be achieved for up to 29 μm [282]. There is ongoing effort to lower the detection energy threshold for μm-wide SNSPDs, to allow access to lighter axion masses.

Additionally, similar to the DCR measurement, the temperature dependence of the PCR measurement was also studied. Figure 9.24 (right) shows the PCR with respect to the bias current, measured at different operating temperature. As shown in the plot, the onset has very little dependence on temperature, compared to the DCR plot. The small dependence observed is due to the fact that at higher temperature, the temperature approaches the critical temperatures, so the energy threshold decreases. Therefore, combining the temperature independence of PCR and the temperature dependence of DCR, allow us to conclude that for DM experiments, it gives us significantly more leverage to operate at the lowest temperature possible.

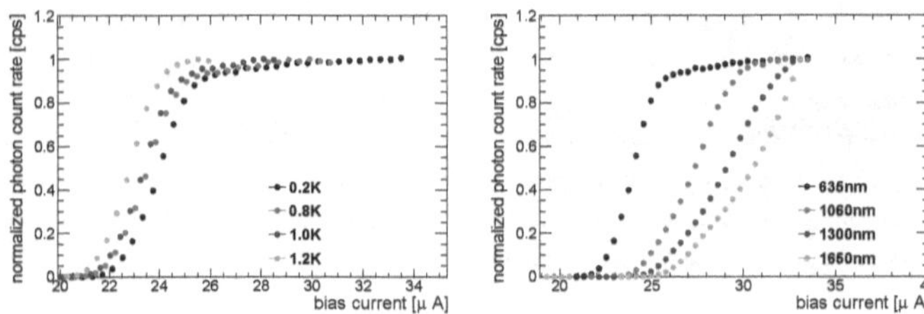

Figure 9.24: The normalized PCR measured at 635 nm for various temperature (left) and normalized PCR measured at 0.2 K for various photon wavelengths (right) are shown.

9.5 Summary and Outlook

In summary, the BREAD experiment proposes a novel experimental design using cylindrical dish resonator with parabolic reflector optimized for embedding in standard solenoids and cryostats. The axion and dark photon target mass extends from $\sim \mu$eV to eV, this large mass range makes it difficult to scale traditional resonator setups to the required volume, while the metallic surfaces can convert DM to photons regardless of their masses. By coupling SNSPDs with the BREAD experiment, we can improve the 0.04–1 eV mass DM reach by several decades. The current progress towards a first stage dark photon pilot experiment with a focus on SNSPD characterization is detailed in this chapter. An unprecedentedly low dark count rate of 10^{-3} cps and a saturated internal detection efficiency for 635 and 1060 nm photon have been measured for the first time for large-area SNSPD. Additionally, the first study on the temperature-dependence of large-area SNSPD performance has been performed. A system to measure the calibrated signal efficiency is being developed for the pilot dark photon experiment.

Finally, the BREAD experiment motivates the sensor development of SNSPD to maximize its potential to detect DM. More light-tight detector package and further understanding of the residual background photons that are observed at the level of 10^{-3} cps would allow us reduce the background rate. Additionally, the photon energy threshold of SNSPD directly determines the axion/dark photon mass sensitivity of the BREAD experiment. Further reducing the energy threshold of large area SNSPDs can by achieved by tuning the silicon content in the wire material to adjust the thermal impedance, as demonstrated for narrower WSi SNSPDs in Ref. [282]. Realizing BREAD as a pioneering DM experiment will catalyze synergies across quantum sensors and particle physics.

10.1   Introduction

Light hidden-sector DM candidates with masses in the MeV-GeV range are theoretically well-motivated and have increasingly been the target of experimental efforts in the last decade [383]. Many hidden-sector dark matter models assume that the basic interaction between the DM and SM particles are through a dark photon portal, which allows the DM to couple to all electrically charged particles. The dark photon-mediated DM are theoretically motivated by many production mechanisms that can realize the DM relic density and give rise to sub-GeV mass scale naturally, including thermal DM with "secluded" annihilation, asymmetric DM [384], and feebly-coupled DM that "freezes in" without reaching equilibrium [385–389]. Given the strong theoretically motivation and little experimental constraints from both the laboratory and astronomical observations, investigating sub-GeV DM is an important and natural direction to pursue.

One of the most sensitive probe for DM is with direct detection experiments, where a DM particle from the Milky-Way halo interacts with a target material producing a detectable signal in the form of heat, phonons, electrons, or photons, captured by a particle detector coupled with the target material [383]. The traditional technique of searching for nuclear recoils loses sensitivity rapidly for DM masses below a few GeV, since the DM is unable to transfer enough of its energy to the nucleus to have any observable signal above detector thresholds. The average energy transferred in an elastic nuclear recoil is $E_{nr} = q^2/2m_N \simeq$ 1 eV $\times (m_{DM}/100\,\mathrm{MeV})^2(10\,\mathrm{GeV}/m_N)$ [390], where m_N is the mass of the nucleus, $q \sim m_{DM}v$ is the momentum transferred, and $v \simeq 10^{-3}$ is the DM velocity. In contrast, if the DM particle scatters off an electron, orders of magnitude lighter than a nucleus, the energy available is significantly larger, $E \simeq m_{DM}v^2 \simeq$ 50 eV $\times (m_{DM}/100\,\mathrm{MeV})$ [390]. This leads to observable signals for DM masses well below 1 GeV, opening up vast new regions of parameters space for experimental exploration.

DM-electron scattering have been investigated in direct DM detection experiments with noble liquid targets and was demonstrated to have sensitivity down to DM masses of a few MeV with an excluded cross section of $\sim 10^{-37}$ cm^2 using XENON10 data [391, 392]. However, the typical noble liquid detectors have a signal threshold of 10 eV when detecting electron ionization [383], which is approximately an order of magnitude higher than that of the semiconductor-based detectors that have a typical band gap energy of a few eV. The semiconductor gallium arsenide (GaAs) is a bright cryogenic scintillator with a low band gap of 1.52 eV, that could potentially probe several orders of magnitude of unexplored DM parameter space for DM masses as low as 1 MeV. Coupling GaAs with an SNSPD, an efficient single photon detector, well-matched to the GaAs emission spectrum, allow us to build a prototype sensing system that can be the basis of new direct DM detection experiments capable of probing masses as low as 1 MeV.

The schematic diagram of the experimental concept is shown in Figure 10.1. Scintillation photons are emitted when an electron excited by a DM-electron scattering interaction relaxes to the ground state [390]. The use of scintillators are preferred over semiconductor ionization detector, such as germanium detectors, because they do not require an electric field, which leads to a more simple setup and potentially fewer dark counts [393]. The SNSPDs then detect the scintillation photons between 0.9 and 1.4 eV with high detection efficiency and low dark count.

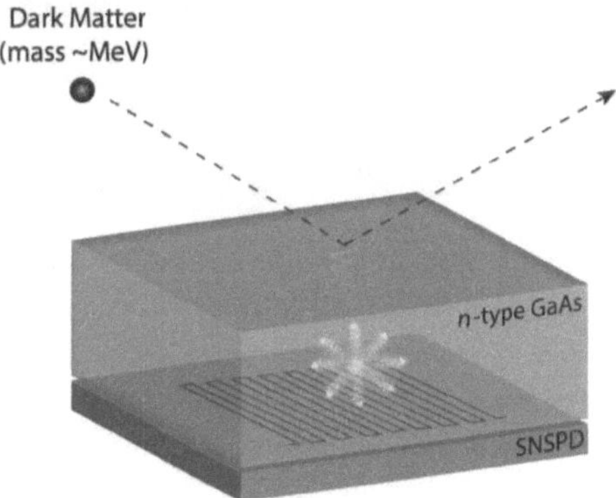

Figure 10.1: An illustration of DM scattering off of an electron in GaAs to produce scintillating photons detected by the SNSPD coupled at the bottom (diagram created by Jamie Luskin).

An n-type GaAs suitably doped with silicon and boron is a cryogenic scintillator with high light yield that makes it an ideal scintillator candidate for MeV-DM searches. Light yields of >100 photons/keV are observed without anti-reflective coatings [394]. The low band gap of 1.52 eV allow us to probe electron excitation from interacting DM as light as 1 MeV [393]. Additionally, there is no afterglow in

GaAs, evidenced by the lack of thermally stimulated luminescence [395]. Finally, GaAs crystals can be commercially produced with high quality and large size of 5 kg [394].

To detect the scintillation photons from GaAs, a highly efficient photon detector for 0.9-1.4 eV photons with low dark count is needed. SNSPDs have demonstrated to have single-photon system detection efficiency as high as 98% at 1550 nm, [396] and dark count rates on the order of 10^{-5} cps. Given these performance metrics, SNSPDs are particularly well-suited to this detection concept. However, unlike the BREAD experiment, where the photons are focused to a focal spot, in this experiment, the signal acceptance is proportional to the target mass of the GaAs and the active area of the SNSPD.

The preliminary sensitivity estimate calculated with the DM-electron scattering rates in GaAs provided by Ref. [393] is shown in Figure 10.2. Two dark photon mediator mass regimes are considered: a heavy mediator with $m_{A'} \gg \alpha m_e$ and a light mediator with $m_{A'} \ll \alpha m_e$, where $m_{A'}$ is the dark photon mass and αm_e is the product of the electron mass and fine structure constant amounting to about 3.7 keV. The corresponding DM form factors are [397]:

$$F_{DM}(q) = \frac{m_{A'}^2 + \alpha^2 m_e^2}{m_{A'}^2 + q^2} \simeq \begin{cases} 1, & m_{A'} \gg \alpha m_e \\ \frac{\alpha^2 m_e^2}{q^2}, & m_{A'} \ll \alpha m_e \end{cases} \tag{10.1}$$

where q is the momentum transfer between the DM and electron. In the sensitivity estimate, zero background events and 100% light collection and photon detection efficiency were assumed. Given the current $\sim$mm^2 active area of SNSPDs, we can build a pathfinder experiment with 1 g of GaAs ($10 \times 10 \times 2\,\text{mm}^3$), coupled to a 1 mm^2 SNSPD, which already allow us to probe regions of the parameter space that have not be explored.

Current efforts on building the pathfinder experiment is described in the rest of the chapter. The progress on characterizing the light yield of the GaAs & SNSPD system with laser excitation, work led by Jamie Luskin (Ph.D. student from University of Maryland/Jet Propulsion Lab), is briefly summarized in Section 10.2. In parallel, I have built and validated an x-ray excitation setup needed to characterize the light yield and energy response of the GaAs & SNSPD system, as described in Section 10.3. Finally, the summary and outlook to larger-scale experiments are presented in Section 10.4.

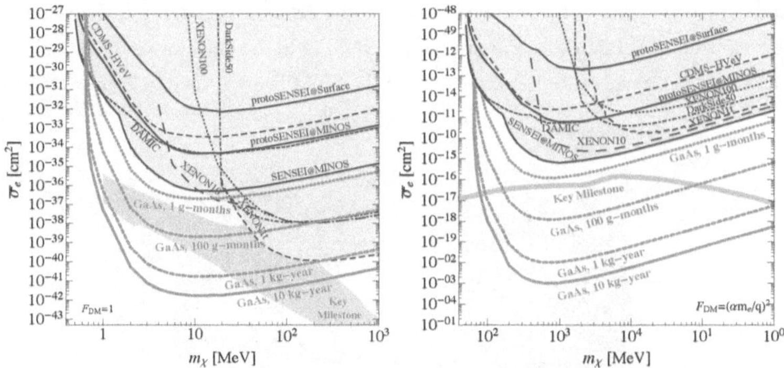

Figure 10.2: Projected sensitivity (blue lines) to dark matter-electron scattering cross section for a heavy mediator (left) or light mediator (right) with a GaAs target. The projected sensitivity assumes zero background events and a light collection and photon detection efficiency of 100%. Existing constraints are shown in gray from SENSEI, DAMIC at SNOLAB, XENON10, XENON100, XENON1T, DarkSide-50, and CDMS HVeV [391, 392, 398–407]. Orange regions labelled "Key Milestone" present a range of model examples in which dark matter obtains the observed relic abundance from its thermal contact with SM particles [408].

10.2 Characterization with Laser

In this section, we briefly summarize the work to measure scintillator decay time and towards light yield measurement with a pulsed 660 nm source. The work described in this section was carried out by Jamie Luskin at Caltech.

The SNSPD that was used for this measurement has the same dimension, fill factor, and detector package as the one used in Section 9.4, but with a slightly different silicon content in the tungsten-silicide wire material. It is an 8-pixel 1×1 mm^2 SNSPD made of tungsten-silicide (42% W and 58% Si). The SNSPD has a fill factor of 0.25 and a critical temperature of 1.9 K. The wires are 1 μm wide and 4.7 nm thick.

The n-type GaAs sample used was sourced from University Wafer. It has a dimension of $5 \times 5 \times 0.625$ mm^3, corresponding to 0.08 g in mass. N-type GaAs samples that have free carrier concentrations from $2 \times 10^{16}/$cm^3 to $6 \times 10^{17}/$cm^3, with silicon and boron concentrations on the order of $1 \times 10^{18}/$cm^3 have been demonstrated to have high light yield. [394]. Therefore, we start with a GaAs sample with $2.99 \times 10^{17}/$cm^3 free carrier concentration, $6.28 \times 10^{17}/$cm^3 boron concentration and $1.14 \times 10^{18} / \text{cm}^3$ silicon concentration. The detector packaging is the same as that of used in Section 9.4. Pictures of the SNSPD in the detector package and the GaAs sample glued onto the active area of the SNSPD are shown in Figure 10.3.

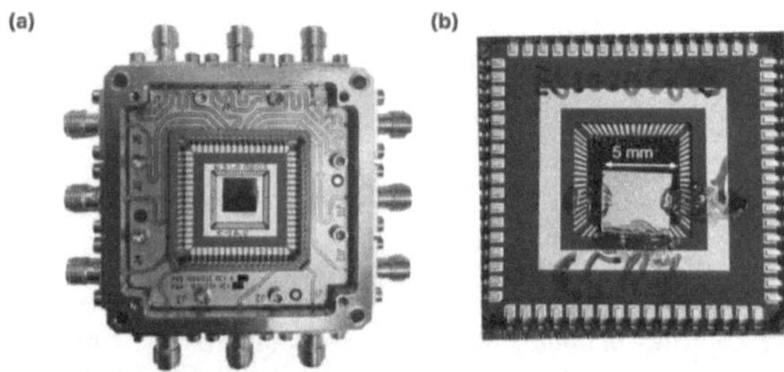

Figure 10.3: (a) The SNSPD and detector packaging used. (b) The SNSPD with GaAs sample glued onto the active area.

To measure the scintillation photons and their decay time in the GaAs due to laser excitation, an attenuated pulsed 660 nm fiber-laser with 10 kHz repetition rate is used to illuminate the GaAs. A 660 nm photons correspond to 1.87 eV, which is just above the 1.52 eV band gap energy for GaAs. The scintillation photons produced by the GaAs is then detected by the SNSPD and readout to room temperature. The GaAs & SNSPD system is operated at 0.8 K temperature and the measurement was done for a single pixel, due to limitations in the available readout lines. The SNSPD signal output along with a synced pulse from the laser are then directed to a room-temperature Swabian Time Tagger, the same type of time-to-digital converter (TDC) used in Chapter 9, allowing us to measure the time difference between the SNSPD signal pulse and the laser. The schematic diagram of the setup is shown in Figure 10.4

The measured scintillator decay time is shown in Figure 10.5. A double exponential function form of $A_1 e^{-\frac{t}{\tau_1}} + A_2 e^{-\frac{t}{\tau_2}} + B$ is fitted to the distribution, resulting in a fast component with $1.2\mu s$ decay time and a slow component with $9.7\mu s$ decay time. The decay time scale is in the same order of magnitude with previous measurements of GaAs decay time using other photon detectors [394]. However, as mentioned in Ref. [394] the decay time depends significantly on the silicon and boron con-

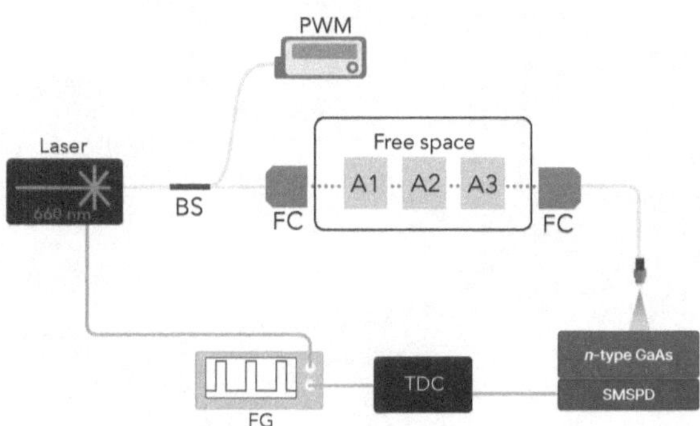

Figure 10.4: The 660 nm laser setup used for optically excited scintillation measurements. The pulsed laser is driven by the function generator (FG) that also sends a synced pulse into a Swabian Time Tagger (TDC). Light is coupled through a 90:10 fiber beam splitter (BS), the output port with 10% of power is diverted to a power meter (PWM) to monitor power fluctuations in the laser. The signal path is attenuated by a series of ND filters (A1-A3) before it is coupled into a fiber that feeds into the fridge to illuminate the GaAs.

centration. Additionally, the measurement that was made was the sum of all four luminescence bands centered at 830, 930, 1070, and 1335 nm [409]. Therefore, the result is not to be compared directly with Ref. [394]. To fully understand the scintillator decay time for each luminescence band would require future study to filter out the photon wavelengths for each band and perform measurements and fits to each band.

In addition to fully understanding the scintillator decay time of each luminescence band, the next step of this project is to measure the light yield of the GaAs & SNSPD system with calibrated laser excitation. Additionally, study to improve the light collection efficiency with different external reflector [410] and to improve the photon absorption efficiency with optical stacks are underway. Once these measurements are understood, the light yield and energy response of the GaAs & SNSPD system will be measured with x-ray excitation, which is described in the following section.

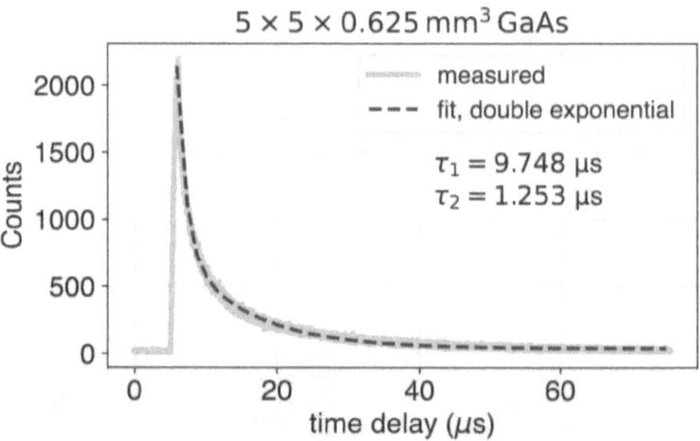

Figure 10.5: The measured time delay between SNSPD signals and the laser is shown. The distribution is fitted to a double exponential function form that resulted in two measured scintillator decay time of 9.7 and $1.2\mu s$.

10.3 Characterization with X-ray

In this section, I will detail the setup and its validation to measure the light yield × detection efficiency of GaAs & SNSPD prototype with x-ray excitation. The setup has been built and validated and is now ready to perform measurements. The x-ray excitation measurements will be performed once the laser excitation measurements are completed. A Compton scattering experiment is setup to "generate" x-ray sources with known energy. In the Compton scattering setup, an incoming 122 keV photon from cobalt-57 (Co-57) scatters off from the GaAs & SNSPD system and produces an outgoing Compton-scattered photon at an angle θ with respect to the original photon detection. The Compton-scattered photon is then detected by a high purity germanium (HP-Ge) detector. HP-Ge detector is used, due to its excellent energy resolution in the 3-300 keV energy range, compared to that of scintillating detectors, such as Sodium Iodide (NaI) detectors. In this setup, only the GaAs & SNSPD system are placed inside of a cryostat. The radioactive source is at room-temperature and the HP-Ge detector requires cooling to cryogenic temperature with liquid nitrogen. A schematic diagram of the experimental setup for the x-ray characterization is shown in Figure 10.6. The energy deposited in the HP-Ge detector (E_{γ_f}) and GaAs ($122\,\mathrm{keV} - E_{\gamma_f}$) are fully determined by the scattering angle

(θ) and the incoming photon energy (E_{γ_i} =122 keV):

$$E_{\gamma_f} = \frac{E_{\gamma_i}}{1 + (E_{\gamma_i}/m_e c^2)(1 - \cos\theta)} \tag{10.2}$$

where m_e is the mass of electron and c is the speed of light. Thus, this setup allow us to measure the number of pixels fired in the SNSPDs, given a known energy deposit.

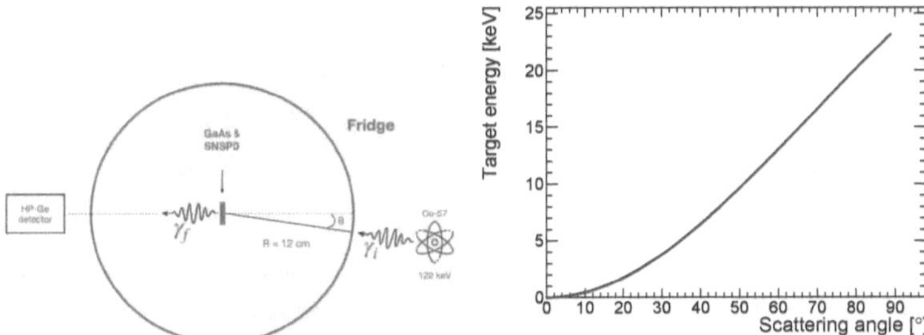

Figure 10.6: A schematic diagram of the Compton scattering setup for the x-ray characterization (left) and the expected energy deposit at the target detector (GaAs & SNSPD) with respect to the scattering angle (right).

The setup was first validated by using two HP-Ge detectors, replacing the GaAs & SNSPD system with another HP-Ge detector, such that we would be able to measure the energy deposited in both detectors. The HP-Ge detectors were first calibrated by measuring the energy spectrum of Co-57. The output signal of the HP-Ge detectors passes through a shaping amplifier that outputs Gaussian pulses with amplitude proportional to the energy. The pulse shape is then sent to an oscilloscope where a histogram of the amplitude of pulses is produced. The amplitude is then converted to energy by matching the peaks in the histogram to four known photon emissions from Co-57 with energies of 6.5, 14.4, 122.1, 136.5 keV. The energy spectrum of Co-57 measured by one of the HP-Ge detectors is shown in Figure 10.7. The four peaks are measured to be at 6.3, 14.2, 121.9, and 136.6 keV with a full width at half maximum (FWHM) of 1.4, 1.9, 1.1, 0.5 keV, agreeing with the specifications from the vendor.

The setup is then validated by setting up the Compton scattering experiment with two HP-Ge detectors located at about 15 cm apart, similar to the radius of the cryostat.

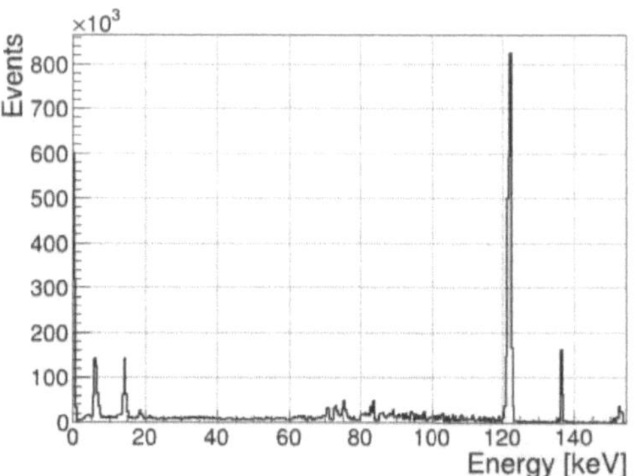

Figure 10.7: Energy spectrum of Co-57 measured by HP-Ge detector.

A picture of the setup is shown in Figure 10.8. Measurements are performed with several angles at 35°, 70°, and 120°, corresponding to target detector energy deposit of 5, 16, and 32 keV, respectively. The measured energy at both detectors agree with the expected energy, calculated based on the scattering angle between the two detectors for all angles. The two-dimensional (2D) energy distribution of the target and tag detectors for the 35° experiment is shown in Figure 10.9 (left). A diagonal line with an energy sum of 122 keV is clearly visible and another fainter diagonal line with a sum of 136 keV is also visible, corresponding to 122 and 136 keV photons from Co-57. The energy distributions of the target and tag detector after requiring the sum of the energy deposit at the detectors to be between 121 and 123 keV are shown in Figure 10.9 (right). Energy peaks of 5.3 ± 1.1 keV and 116.3 ± 1.3 keV are measured at the target and tag detector, respectively, agreeing with the expected energy of 5 and 117 keV for a scattering angle of 35°. Similar agreements were observed for scattering angles 70° and 120°. Therefore, we have proved that we can deposit x-ray energy between 5 ($\theta = 35°$) and 32 keV ($\theta = 120°$) in the GaAs.

Additionally, mechanical work to integrate and align the GaAs, radioactive source, and HP-Ge detector has also been completed. The HP-Ge detector has been raised to the height of the SNSPD in the cryostat with an 80/20 aluminum stage. An automated stage that can align the Co-57 source to the SNSPD and scan multiple scattering angles has been setup, as shown in Figure 10.10. The system is currently ready to perform the x-ray characterization and measurement will be taken as soon as the work described in Section 10.2 concludes.

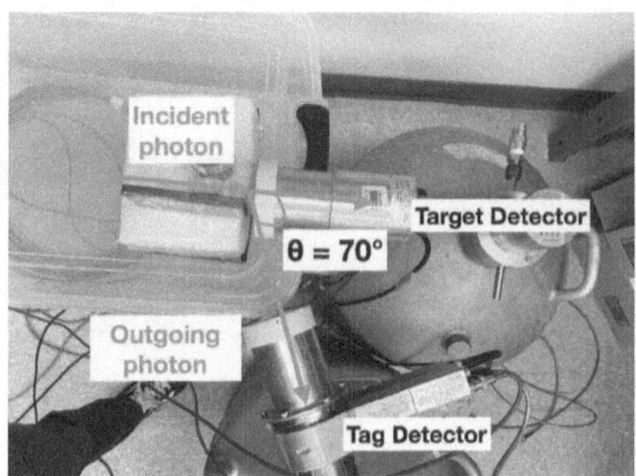

Figure 10.8: A picture of the setup to validate the Compton scattering experiment with two HP-Ge detectors is shown. The radioactive source is placed as close as possible to the target detector, where the Compton scattering occurs, to maximize the rate. Lead plates are placed next to the source to block any photons from Co-57 to directly reach the tag detector. The tag detector is placed at an angle of $70°$ from the target detector.

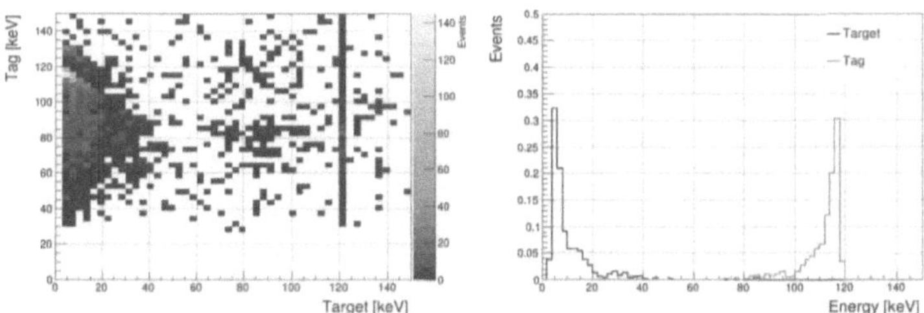

Figure 10.9: The 2D energy distribution (left) of the tag and target detector and the individual energy distributions (right) of the two detectors after requiring the their energy sum to be between 121 and 123 keV. The energy distributions correspond to a scattering angle of $35°$.

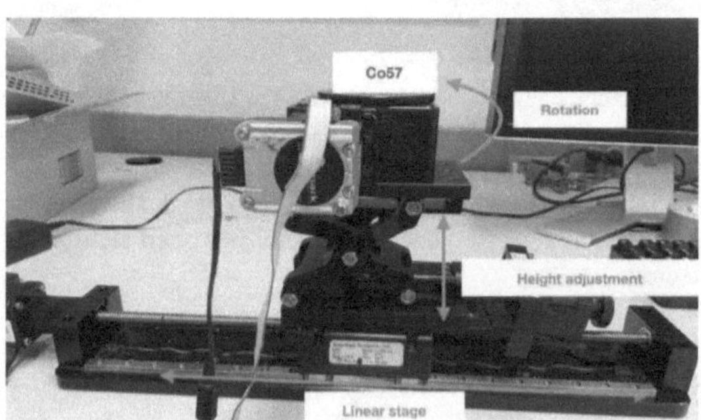

Figure 10.10: A picture of the automated stages that align the radioactive source with the SNSPDs at different angles.

10.4 Summary and Outlook

In summary, we have proposed a novel experimental design using n-type GaAs target that has a low band gap energy of 1.52 eV and produces 0.9-1.4 eV scintillation photons. These photons can be efficiently detected by SNSPDs with single-photon sensitivity. In particular, little background is expected, since the GaAs target has no afterglow and the scintillation photons can be easily measured without the need of an amplification mechanism (external electric field). This detection concept could be successfully realized with a large mass target (> 10 kg) and improve the reach to DM-electron cross section by orders of magnitude for MeV-DM.

Given the current mm^2 active area of SNSPD, we can build a pathfinder experiment with 1 g of target GaAs coupled to a 1 mm^2 SNSPD, that can already improve the sensitivity reach to DM. The current progress towards the pathfinder experiment on the characterization of the GaAs & SNSPD system with optical and x-ray excitations was summarized in this chapter. The pathfinder experiment will be commissioned, once a complete and thorough understanding of the light yield and energy response of the system is obtained. In parallel, there are also ongoing efforts to study the background by building scintillators around the cryostat to study and remove cosmic muon tracks that may penetrate the target and by studying the time correlation of events across pixels for background events.

Scaling this experiment to larger target volumes drives the development of large-area SNSPDs with cm^2-scale active areas. Recent advances in the preparation of thin superconducting films have allowed the fabrication of SNSPDs with μm-scale wire widths. These devices are amenable to fabrication with photolithography [264, 411], a more robust and scalable manufacturing technique than the traditional electron beam lithography used for SNSPDs with nm-scale wires. To achieve kg-scale target masses, 100-pixel SNSPD modules will be used, requiring scalable multiplexing readout system. Frequency-domain multiplexing techniques that have traditionally been implemented in other quantum detectors have recently been demonstrated on a 16-pixel SNSPD array that can be readout on a single microwave line [412]. By combining the SNSPDs with such frequency multiplexing schemes, sufficiently large SNSPDs can be realized.